CRISPR GENOME SURGERY IN STEM CELLS AND DISEASE TISSUES

CRISPR GENOME SURGERY IN STEM CELLS AND DISEASE TISSUES

Edited by

STEPHEN H. TSANG

Departments of Ophthalmology, Biomedical Engineering, Pathology & Cell Biology; Graduate Programs in Nutritional & Metabolic Biology and Neurobiology & Behavior, Columbia Stem Cell Initiative, New York, NY, United States

ELSEVIER

ACADEMIC PRESS

An imprint of Elsevier

Academic Press is an imprint of Elsevier
125 London Wall, London EC2Y 5AS, United Kingdom
525 B Street, Suite 1650, San Diego, CA 92101, United States
50 Hampshire Street, 5th Floor, Cambridge, MA 02139, United States
The Boulevard, Langford Lane, Kidlington, Oxford OX5 1GB, United Kingdom

British Library Cataloguing-in-Publication Data
A catalogue record for this book is available from the British Library

Library of Congress Cataloging-in-Publication Data
A catalog record for this book is available from the Library of Congress

ISBN: 978-0-12-817876-8

For Information on all Academic Press publications
visit our website at https://www.elsevier.com/books-and-journals

Publisher: Stacy Masucci
Acquisitions Editor: Elizabeth Brown
Editorial Project Manager: Pat Gonzalez
Production Project Manager: Omer Mukthar
Cover Designer: Victoria Pearson

Typeset by MPS Limited, Chennai, India

Contents

6. Targeting Alzheimer's disease and related dementias with CRISPR and human pluripotent stem cell technologies 65

ANDREW A. SPROUL

7. Gene editing for the cornea 81

M. ANDREW NESBIT AND C.B. TARA MOORE

8. Employing nonhomologous end joining and homology-directed repair for treatment of Leber congenital amaurosis and inherited retinal degeneration 101

DONG HYUN JO, JIN HYOUNG KIM AND JEONG HUN KIM

9. The ethics of gene editing in human stem cells 111

ALANGOYA TEZEL

List of contributors

Angela Y. Chang Edward S. Harkness Eye Institute, Department of Ophthalmology, Columbia University Irving Medical Center, New York, NY, United States

Dong Hyun Jo Department of Anatomy and Cell Biology, Seoul National University College of Medicine, Seoul, Republic of Korea

Jeong Hun Kim Fight against Angiogenesis-Related Blindness (FARB) Laboratory, Biomedical Research Institute, Seoul National University Hospital, Seoul, Republic of Korea; Department of Biomedical Sciences, Seoul National University College of Medicine, Seoul, Republic of Korea; Department of Ophthalmology, Seoul National University College of Medicine, Seoul, Republic of Korea

Jin Hyoung Kim Fight against Angiogenesis-Related Blindness (FARB) Laboratory, Biomedical Research Institute, Seoul National University Hospital, Seoul, Republic of Korea

C.B. Tara Moore Biomedical Sciences Research Institute, Ulster University, Coleraine, United Kingdom

M. Andrew Nesbit Biomedical Sciences Research Institute, Ulster University, Coleraine, United Kingdom

Margaret E. Scollan Columbia University College of Physicians and Surgeons, New York, NY, United States

Falak Sher Center for Translational & Computational Neuroimmunology, Taub Institute for Research on Alzheimer's Disease and the Aging Brain, Department of Neurology, Columbia University Medical Center, Columbia University, New York, NY, United States

Andrew A. Sproul Taub Institute for Research on Alzheimer's Disease and the Aging Brain, Columbia University, New York, NY, United States; Department of Pathology and Cell Biology, Columbia University Irving Medical Center, New York, NY, United States

Alangoya Tezel University of Michigan Medical School, Ann Arbor, MI, United States

Stephen H. Tsang Department of Ophthalmology, Columbia University, New York, NY, United States; Jonas Children's Vision Care, Bernard & Shirlee Brown Glaucoma Laboratory, Columbia University, New York, NY, United States; Department of Pathology & Cell Biology, Columbia University, New York, NY, United States; Institute of Human Nutrition, College of Physicians and Surgeons, Columbia University, New York, NY, United States

Christine L. Xu Department of Ophthalmology, Stanford University, Stanford, CA, United States; Department of Ophthalmology, Columbia University, New York, NY, United States; Jonas Children's Vision Care, Bernard & Shirlee Brown Glaucoma Laboratory, Columbia University, New York, NY, United States

Foreword

New advances in gene editing and precision genome surgery have paved the way for therapies in regenerative medicine. Therapeutic gene-editing technology is constantly evolving. This book outlines recent and important innovations in clustered-regularly interspaced short palindromic repeats (CRISPR)/CRISPR-associated (Cas) technology that is now at the forefront of precision medicine and engineering of stem cells. Such technology holds great promise for advancing new scientific discoveries and for bench-to-bedside translation in the near future.

This book provides an overview of the history of CRISPR technology, CRISPR-Cas orthologues, CRISPR genome engineering techniques, and applications of gene-editing technologies to inherited disorders. Examples of translational applications holding great promises include Alzheimer's disease and inherited retinal dystrophies, a diverse group of progressive neurodegenerative disorders. This book concludes with an overview of the ethics of gene editing in human stem cells, to tie together how such important scientific advances affect society today.

An important read for anyone interested in understanding the power and challenges of human gene editing technologies for regenerative medicine.

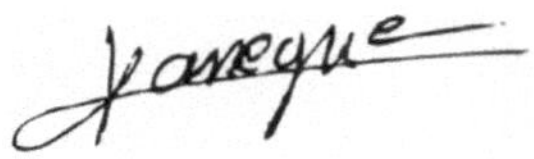

Prof. Emmanuelle Passegué[1,2]

[1]*Director of the Columbia Stem Cell Initiative, Columbia University Irving Medical Center, New York, NY, United States* [2]*Endowed Alumni Professor, Department of Genetics and Development, Columbia University Irving Medical Center, New York, NY, United States*

CHAPTER

1

The history of CRISPR: from discovery to the present

Christine L. Xu[1,2,3] *and Stephen H. Tsang*[2,3,4,5]

[1]Department of Ophthalmology, Stanford University, Stanford, CA, United States [2]Department of Ophthalmology, Columbia University, New York, NY, United States [3]Jonas Children's Vision Care, Bernard & Shirlee Brown Glaucoma Laboratory, Columbia University, New York, NY, United States [4]Department of Pathology & Cell Biology, Columbia University, New York, NY, United States [5]Institute of Human Nutrition, College of Physicians and Surgeons, Columbia University, New York, NY, United States

The beginnings of CRISPR/Cas9

The history of clustered regularly interspaced short palindromic repeats (CRISPR)/Cas9 began [1] with a microbiologist by the name of Francisco Mojica, who worked at the University of Alicante in Spain [2]. His research centered around a microbe called *Haloferax mediterranei*, a species in the archaea family known for its tolerance to extreme salt levels. This tolerance was achieved because restriction enzymes cut the microbe's genome in a manner that was dependent on the surrounding salt levels. In 1993, Mojica et al. discovered a unique characteristic in the genome of these *H. mediterranei*: repeat sequences of 30 base pairs (bp) separated by "spacer" sequences of roughly 36 bp. He was surprised to find that these spacers did not look like typical microbial sequences of DNA [3].

Upon further investigation, Mojica et al. discovered that these curious repeat sequences were preserved in other species as well, such as *Haloferax volcanii* and the more distantly related *Halophilic archaea* [3]. Ishino et al. reported similar findings in 1987, discovering repeat sequences in *Escherichia coli* [4]. Over the following 20 years, Mojica et al. would find loci for "short regularly spaced repeats"—which would later be renamed to CRISPR—in over 20 different microbial species [5–7]. Using his Basic Local Alignment Search Tool program, he analyzed spacers from different species and determined whether they had genetic similarity to any known DNA sequences. He found that roughly two-thirds of the 88 analyzed microbial spacers matched to viruses or conjugative plasmids related to the microbes. From this data, he hypothesized that CRISPR loci might be involved in the

CRISPR Genome Surgery in Stem Cells and Disease Tissues
DOI: https://doi.org/10.1016/B978-0-12-817876-8.00007-3

microbial adaptive immune system as a protective mechanism against pathogens [8].

Further investigations by Gilles Vergnaud, a geneticist who analyzed *Yersinia pestis* strains, led to the discovery that their tandem repeat loci were all identical, except for one distinct difference at a "CRISPR locus"; this particular locus had spacers matching the DNA of a prophage in the *Y. pestis* genome. As a result, he hypothesized that the CRISPR locus was influential in immunological defense mechanisms by retaining information of past pathogens and aggressors in their "genetic memory" [9].

Underlying mechanisms of CRISPR/Cas9

Subsequent CRISPR research aimed to reveal the exact manner in which this mechanism was orchestrated. Alexander Bolotin suggested that transcripts from the CRISPR locus inhibited the phage gene expression via antisense RNA inhibition [10]. This was later disproved by Barrangou et al., a group that analyzed the *Streptococcus thermophilus* strain and two bacteriophages [10–12]. They isolated the phage-resistant bacteria in their sample and found that the bacteria all shared the phage-derived sequences incorporated into their CRISPR locus, thereby confirming the mechanism by which CRISPR bestowed adaptive immunity to bacteria [13].

John Van der Oost devoted much of his research to the discovery of the specific molecular mechanisms involved in the CRISPR program [10]. In a 2008 study published in *Science*, Brouns et al. reported that five Cas proteins (collectively called "Cascade") were necessary for cleaving a long precursor RNA strand derived from the CRISPR locus into a compact CRISPR RNA (crRNA) strand [14,15]. Furthermore, palindromic repeats were determined to help in the process of building the crRNA's secondary structure. CRISPR was also programmed to target genes in the lambda phage, marking the first time in history that CRISPR was successfully programmed. The study concluded with the hypothesis that CRISPR targets DNA rather than RNA.

In 2008, Marraffini and Sontheimer provided conclusive evidence to support this hypothesis [16]. Through studying *Staphylococcus epidermis*, Luciano Marraffini et al. found that one of the bacterial spacers matched a nickase gene present in antibiotic-resistant *Staphylococcus aureus* plasmids. Plasmids normally can undergo horizontal gene transfer in bacteria and archaea via phage transduction, transformation, or conjugation. In this case, the staphylococcal conjugative plasmids could not undergo horizontal gene transfer in *S. epidermis*. When destroying the function of the nickase gene or the spacers in the *S. epidermis* CRISPR locus, however, interference no longer occurred. This suggested that CRISPR must be the causal mechanism responsible for blocking the plasmids from conjugation or plasmid transformation.

Subsequent research performed by Deveau et al. and Horvath et al. in 2008 reported more data on the mechanisms of CRISPR genome editing [17,18]. Direct observations of CRISPR in action in the *S. thermophilus* strain revealed that the Cas9 nuclease performed the plasmid cutting. Specifically, the cuts were made precisely at a location three nucleotides upstream of the "protospacer adjacent motif" (PAM). Viral DNA was also shown to be cut at the same location. Their research ultimately showed that Cas9 within the CRISPR system was responsible for making cuts at specific locations designated by the guiding crRNA sequence.

Emmanuelle Charpentier's research revealed an additional component in the CRISPR/Cas9 machinery that worked in conjunction with crRNA [19]. Through parallel sequencing of the RNA from *Streptococcus pyogenes*, Dr. Charpentier's team discovered numerous novel small RNA sequences (~24 nucleotides) with complementarity to CRISPR repeats. This small trans-activating RNA (tracrRNA) was transcribed from a section of DNA adjacent to the CRISPR locus. Deleting tracrRNA caused the CRISPR system to become completely nonoperational. Therefore, Charpentier et al. discovered that the tracrRNA and crRNA hybridization and consequent cleavage by RNaseIII into a mature final product is an important step that allows the Cas9 protein to find the designated target sequence.

Virginijus Siksnys, a chemist, and his collaborators demonstrated that transferring the CRISPR locus from *S. thermophilus* into *E. coli* essentially gave *E. coli* the tools for adaptive immunity [20]. Specifically, *E. coli* could use the CRISPR locus to perform specific interference against plasmid and bacteriophage DNA, giving it protection from invaders. Furthermore, investigators discovered that the RuvC and HNH nuclease domains were responsible for making the double-stranded cuts in DNA [21,22].

Later on, Siksnys and collaborators researched CRISPR/Cas9 in a test tube [23]. In these studies, they purified the Cas9 and crRNA from *S. thermophilus* by tagging Cas9 with streptavidin in vivo and found that the Cas9 was able to cleave DNA in vitro in the same way that it functioned in vivo (making a double-stranded cut three nucleotides away from the PAM sequence). Furthermore, the researchers were able to use CRISPR to cleave a designated target site, specified by a customized spacer in the CRISPR locus. Mutating the two nuclease domains in Cas9 also demonstrated that the HNH domain is responsible for cleaving the strand complementary to the crRNA and that the RuvC domain is responsible for cleaving the opposite strand.

In 2012, Jinek et al. reported that a single guide RNA (sgRNA)—in other words, an engineered single RNA chimera—was able to carry out the same function as the tracrRNA:crRNA complex [24]. This breakthrough would change the future of CRISPR by making it more efficient and programmable for research purposes. Jinek et al.'s keen observation of this system's great potential "to exploit the [CRISPR] system for RNA-programmable genome editing" foreshadows the future of genome engineering.

Using CRISPR/Cas9 for mammalian gene editing

In 2013, multiple labs demonstrated successful CRISPR gene editing in mammalian cells. Cong et al. published a paper in *Science* reporting their design of tracrRNA and pre-crRNA, used to target the human *EMX1* locus in mammalian cells together with *S. pyogenes* Cas9 (SpCas9) [25]. This team of researchers also developed a crRNA:tracrRNA duplex with chimeric RNA, and targeted designated genomic locations in human and mouse cells (human *PVALB* and mouse *Th*). George Church's laboratory explored CRISPR's utility in both nonhomologous end joining (NHEJ) and homologous recombination in mammalian cells [26]. Jennifer Doudna's laboratory published results of CRISPR gene editing in human cells using sgRNA [27]. Keith Joung's laboratory reported targeting zebrafish embryos in vivo, using CRISPR to introduce deletions into the germline.

Future directions

It is difficult to imagine that the bacterial protective mechanism against pathogens—with its spacer sequences and curious palindromic repeats encoded into the genome—was discovered by Mojica in 1993, less than 30 years prior to the publication of this book [2]. With the large amount of research devoted to CRISPR/Cas9 experiments today, it is even harder to believe that the first CRISPR/Cas9 application in mammalian gene editing occurred only 8 years prior to this book.

Before translating the CRISPR/Cas9 genome editing tool from animal studies into clinical trials, its safety and efficacy must first be established. One of the main concerns regarding CRISPR/Cas9 technology is off-target mutagenesis, or mutations caused in locations other than the designated target locus [28]. The sgRNA used to guide Cas9 to the target site is usually 20 nucleotides long, and off-target effects could occur with even as many as 3–5 bp mismatches in the PAM distal part of the sgRNA [25,29,30]. As a result, many tools are being developed to minimize off-target effects without sacrificing CRISPR editing efficiency. Cas9 from *Streptococcus aureus* has less restrictions on PAM specificity, and also demonstrated high efficacy and specificity in human cells [31,32]. *Campylobacter jejuni* Cas9 (CjCas9) is a small Cas9 orthologue only 2.95 kb long, significantly smaller than SpCas9 (4.2 kb) [32,33]. Besides the benefits of this small size, which allows CjCas9 to fit comfortably in adeno-associated viruses (which have carrying capacities of 4.2 kb), they also have been shown to have high specificity and editing efficiency. Of equal interest is Cas12a from *Lachnospiraceae bacterium ND2006* (LbCpf1) and *Acidaminococcus* sp. *BV3L6* (AsCpf1) due to their high targeting specificity and optimal NHEJ-directed repair, achieved from creating staggered cuts with "sticky ends" rather than blunt double end cuts like SpCas9 [32,34,35].

New platforms have also been developed to analyze off-target mutations in CRISPR-edited samples. In order to obtain a genome-wide, unbiased identification of double-stranded breaks (DSBs), a GUIDE-seq sequencing tool conducts a genome-wide analysis of the off-targeting effects using the global capture of DSBs created by Cas9 [36]. High-throughput, genome-wide translocation sequencing analyzes off-target cleavage by way of translocations created by "bait" DSBs at target sites and "prey" DSBs throughout the genome [37]. In vitro Cas9-digested whole-genome sequencing offers a robust, unbiased, and cost-effective way to track off-target mutations based on the sequence alignment of DNA fragments with identical versus staggered 5′ ends [38]. Finally, breaks labeling in situ and sequencing is a platform that uses linear amplification of DSBs in fixed cells to efficiently analyze Cas9-generated off-target mutations [39].

Some researchers have focused on harnessing CRISPR in other ways. A method of CRISPR interference has been generated by using catalytically dead Cas9 (dCas9) to bind to the target locus and effectively block RNA polymerase from binding without generating any DSBs [40]. Mutating just one of the two catalytic sites in Cas9 creates DNA nickases, which make single-stranded cuts in the genome and generate lower off-target effects [23,41,42]. David Liu has dedicated much of his research to creating CRISPR base editors, which use cytidine or adenosine deaminases to change single bases in the genome. For example, fusing apolipoprotein B mRNA editing enzyme complex-1 cytidine deaminase to dCas9 will convert a C:G base pair to a T:A base pair [43].

CRISPR may be a significant genome editing tool for human therapies in the future, but before they can be translated from bench to bedside, strict regulations need to be put into place. Jennifer Doudna has frequently warned the public on the ethics of CRISPR editing in humans, emphasizing that the cost-effective and relative ease of using CRISPR (i.e., "you don't have to have money, or a lot of connections ... to get a hold [of the technology]") should increase the level of precautions surrounding the use of CRISPR in humans [44]. When He Jiankui reported his use of CRISPR to edit the *CCR5* gene in two embryos (later born as twin girls), the public outcry was immediate [1]. As CRISPR editing in humans has become an increasingly likely possibility in today's world, it is imperative that stricter global regulations be implemented. By establishing such guidelines, scientists would gain a greater understanding on the ethical implications involved in genome editing, as well as how to deliver genetic engineering in the most safe and efficacious way. These steps are critical prior to using CRISPR/Cas9 technology in gene therapies and direct human germline editing.

Acknowledgments

This study supported, in part, by The Jonas Children's Vision Care and Bernard & Shirlee Brown Glaucoma Laboratory are supported by the National Institutes of Health (P30EY019007, R01EY018213, §R01EY024698, R01EY026682, R21AG050437), National Cancer Institute Core (5P30CA013696), Foundation Fighting Blindness (TA-NMT-0116-0692-COLU), the Research to Prevent Blindness (RPB) Physician-Scientist Award, and unrestricted funds from RPB, New York, NY, United States. SHT is a member of the RD-CURE Consortium and is supported by Kobi and Nancy Karp, the Crowley Family Fund, the Rosenbaum Family Foundation, the Tistou and Charlotte Kerstan Foundation, the Schneeweiss Stem Cell Fund, New York State (C029572), and the Gebroe Family Foundation.

Author contributions

CLX wrote and edited the manuscript. SHT oversaw the writing process.

Conflict of interest statement

The authors declare that the research was conducted in the absence of any commercial or financial relationships that could be construed as a potential conflict of interest.

References

[1] D. Cyranoski, The CRISPR-baby scandal: what's next for human gene-editing, Nature 566 (7745) (2019) 440–442.

[2] F.J. Mojica, G. Juez, F. Rodriguez-Valera, Transcription at different salinities of *Haloferax mediterranei* sequences adjacent to partially modified PstI sites, Mol. Microbiol. 9 (3) (1993) 613–621.

[3] F.J. Mojica, et al., Long stretches of short tandem repeats are present in the largest replicons of the Archaea *Haloferax mediterranei* and *Haloferax volcanii* and could be involved in replicon partitioning, Mol. Microbiol. 17 (1) (1995) 85–93.

[4] Y. Ishino, et al., Nucleotide sequence of the iap gene, responsible for alkaline phosphatase isozyme conversion in *Escherichia coli*, and identification of the gene product, J. Bacteriol. 169 (12) (1987) 5429–5433.

[5] R. Jansen, et al., Identification of genes that are associated with DNA repeats in prokaryotes, Mol. Microbiol. 43 (6) (2002) 1565–1575.

[6] F.J. Mojica, R.A. Garrett, Discovery and seminal developments in the CRISPR field, in: R. Barrangou, J. van der Oost (Eds.), CRISPR-Cas Systems, Springer, Berlin/Heidelberg, Germany, 2013, pp. 1–31.

[7] F.J. Mojica, et al., Biological significance of a family of regularly spaced repeats in the genomes of Archaea, Bacteria and mitochondria, Mol. Microbiol. 36 (1) (2000) 244–246.

[8] F.J. Mojica, et al., Intervening sequences of regularly spaced prokaryotic repeats derive from foreign genetic elements, J. Mol. Evol. 60 (2) (2005) 174–182.

[9] C. Pourcel, G. Salvignol, G. Vergnaud, CRISPR elements in *Yersinia pestis* acquire new repeats by preferential uptake of bacteriophage DNA, and provide additional tools for evolutionary studies, Microbiology 151 (Pt 3) (2005) 653–663.

[10] E.S. Lander, The heroes of CRISPR, Cell 164 (1-2) (2016) 18–28.

[11] R. Barrangou, CRISPR-Cas systems and RNA-guided interference, Wiley Interdiscip. Rev. RNA 4 (3) (2013) 267–278.

[12] R. Barrangou, et al., Advances in CRISPR-Cas9 genome engineering: lessons learned from RNA interference, Nucleic Acids Res. 43 (7) (2015) 3407–3419.

[13] R. Barrangou, et al., CRISPR provides acquired resistance against viruses in prokaryotes, Science 315 (5819) (2007) 1709–1712.

[14] S.J. Brouns, et al., Small CRISPR RNAs guide antiviral defense in prokaryotes, Science 321 (5891) (2008) 960–964.

[15] R. Sorek, V. Kunin, P. Hugenholtz, CRISPR--a widespread system that provides acquired resistance against phages in bacteria and archaea, Nat. Rev. Microbiol. 6 (3) (2008) 181–186.

[16] L.A. Marraffini, E.J. Sontheimer, CRISPR interference limits horizontal gene transfer in *staphylococci* by targeting DNA, Science 322 (5909) (2008) 1843–1845.

[17] P. Horvath, et al., Diversity, activity, and evolution of CRISPR loci in *Streptococcus thermophilus*, J. Bacteriol. 190 (4) (2008) 1401–1412.

[18] H. Deveau, et al., Phage response to CRISPR-encoded resistance in *Streptococcus thermophilus*, J. Bacteriol. 190 (4) (2008) 1390–1400.

[19] E. Deltcheva, et al., CRISPR RNA maturation by trans-encoded small RNA and host factor RNase III, Nature 471 (7340) (2011) 602–607.

[20] R. Sapranauskas, et al., The *Streptococcus thermophilus* CRISPR/Cas system provides immunity in *Escherichia coli*, Nucleic Acids Res. 39 (21) (2011) 9275–9282.

[21] A. Bolotin, et al., Clustered regularly interspaced short palindrome repeats (CRISPRs) have spacers of extrachromosomal origin, Microbiology 151 (Pt 8) (2005) 2551–2561.

[22] K.S. Makarova, et al., A putative RNA-interference-based immune system in prokaryotes: computational analysis of the predicted enzymatic machinery, functional analogies with eukaryotic RNAi, and hypothetical mechanisms of action, Biol. Direct 1 (2006) 7.

[23] G. Gasiunas, et al., Cas9-crRNA ribonucleoprotein complex mediates specific DNA cleavage for adaptive immunity in bacteria, Proc. Natl Acad. Sci. U.S.A. 109 (39) (2012) E2579–E2586.

[24] M. Jinek, et al., A programmable dual-RNA-guided DNA endonuclease in adaptive bacterial immunity, Science 337 (6096) (2012) 816–821.

[25] L. Cong, et al., Multiplex genome engineering using CRISPR/Cas systems, Science 339 (6121) (2013) 819–823.

[26] P. Mali, et al., RNA-guided human genome engineering via Cas9, Science 339 (6121) (2013) 823–826.

[27] M. Jinek, et al., RNA-programmed genome editing in human cells, eLife 2 (2013) e00471.

[28] M. Jamal, et al., Keeping CRISPR/Cas on-target, Curr. Issues Mol. Biol. 20 (2016) 1–12.

[29] V. Pattanayak, et al., High-throughput profiling of off-target DNA cleavage reveals RNA-programmed Cas9 nuclease specificity, Nat. Biotechnol. 31 (9) (2013) 839–843.

[30] Y. Fu, et al., High-frequency off-target mutagenesis induced by CRISPR-Cas nucleases in human cells, Nat. Biotechnol. 31 (9) (2013) 822–826.

[31] B.P. Kleinstiver, et al., Engineered CRISPR-Cas9 nucleases with altered PAM specificities, Nature 523 (7561) (2015) 481–485.

[32] C.L. Xu, et al., Viral delivery systems for CRISPR, Viruses 11 (1) (2019).

[33] E. Kim, et al., In vivo genome editing with a small Cas9 orthologue derived from *Campylobacter jejuni*, Nat. Commun. 8 (2017) 14500.

[34] T. Yamano, et al., Structural basis for the canonical and non-canonical PAM recognition by CRISPR-Cpf1, Mol. Cell 67 (4) (2017) 633–645.

[35] D. Kim, et al., Genome-wide analysis reveals specificities of Cpf1 endonucleases in human cells, Nat. Biotechnol. 34 (8) (2016) 863–868.

[36] S.Q. Tsai, et al., GUIDE-seq enables genome-wide profiling of off-target cleavage by CRISPR-Cas nucleases, Nat. Biotechnol. 33 (2) (2015) 187–197.

[37] R.L. Frock, et al., Genome-wide detection of DNA double-stranded breaks induced by engineered nucleases, Nat. Biotechnol. 33 (2) (2015) 179–186.

[38] D. Kim, et al., Digenome-seq: genome-wide profiling of CRISPR-Cas9 off-target effects in human cells, Nat. Methods 12 (3) (2015) 237–243.

[39] R. Mirzazadeh, et al., Genome-wide profiling of DNA double-strand breaks by the BLESS and BLISS methods, Method. Mol. Biol. 1672 (2018) 167–194.

[40] L.S. Qi, et al., Repurposing CRISPR as an RNA-guided platform for sequence-specific control of gene expression, Cell 152 (5) (2013) 1173–1183.

[41] P. Mali, et al., CAS9 transcriptional activators for target specificity screening and paired nickases for cooperative genome engineering, Nat. Biotechnol. 31 (9) (2013) 833–838.

[42] F.A. Ran, et al., Double nicking by RNA-guided CRISPR Cas9 for enhanced genome editing specificity, Cell 154 (6) (2013) 1380–1389.

[43] H.A. Rees, D.R. Liu, Base editing: precision chemistry on the genome and transcriptome of living cells, Nat. Rev. Genet. 19 (12) (2018) 770–788.

[44] E. Shao, Y. Pershad, CRISPR co-inventor Jennifer Doudna talks ethics and biological frontiers. <https://ethicsinsociety.stanford.edu/>, 2019 (accessed 19.04.02).

CHAPTER

2

CRISPR-Cas orthologs and variants

Margaret E. Scollan

Columbia University College of Physicians and Surgeons, New York, NY, United States

Introduction

Since the discovery and initial application of *Streptococcus pyogenes* Cas9 (SpCas9) for gene editing, a wealth of structural and functional data regarding clustered regularly interspaced short palindromic repeats (CRISPR)-Cas systems has come to light. SpCas9 is continuously being optimized for improved delivery, activity, and target recognition while simultaneously being reimagined for a range of applications. Moreover, the CRISPR-Cas toolbox is expanding as Cas orthologs, and related effector proteins from other bacterial species are discovered and remodeled. This chapter seeks to describe the diversity of the CRISPR-Cas systems by describing the SpCas9 variants that have been engineered to improve specificity and efficiency, investigating orthologous Cas9 variants and related nucleases, and expounding on the variety of CRISPR-Cas applications. This information will prove useful in the scientific communities' continued efforts to optimize gene-editing protocols and engineer therapeutic strategies.

Cas9: a model of CRISPR-Cas systems

CRISPR systems are involved in the prokaryotic protective response to foreign material. They are characterized by an RNA-guided, "CRISPR-associated" nuclease (Cas) that recognizes and cleaves foreign DNA. Known CRISPR-Cas systems have been divided into 2 major classes and further split into 5 types and 16 subtypes [1]. Class 1 types (I, III, and IV) consist of a multisubunit effector complex while Class 2 types (II, V, VI) have a single effector protein (Fig. 2.1) [1]. Class 2 systems are more commonly utilized for gene editing due to their simplicity.

The most widely known Class 2 type II system is characterized by its signature nuclease, Cas9, which introduces double-stranded breaks (DSBs) using two catalytic residues, RuvC and HNH, to respectively nick opposite strands of DNA [4,5]. In nature, Cas is guided to its target by two small RNAs, a CRISPR RNA (crRNA) and transactivating CRISPR RNA (tracrRNA). Jinek et al. streamlined this system by redesigning the crRNA:tracrRNA complex into a single "guide RNA" transcript

CRISPR Genome Surgery in Stem Cells and Disease Tissues
DOI: https://doi.org/10.1016/B978-0-12-817876-8.00008-5

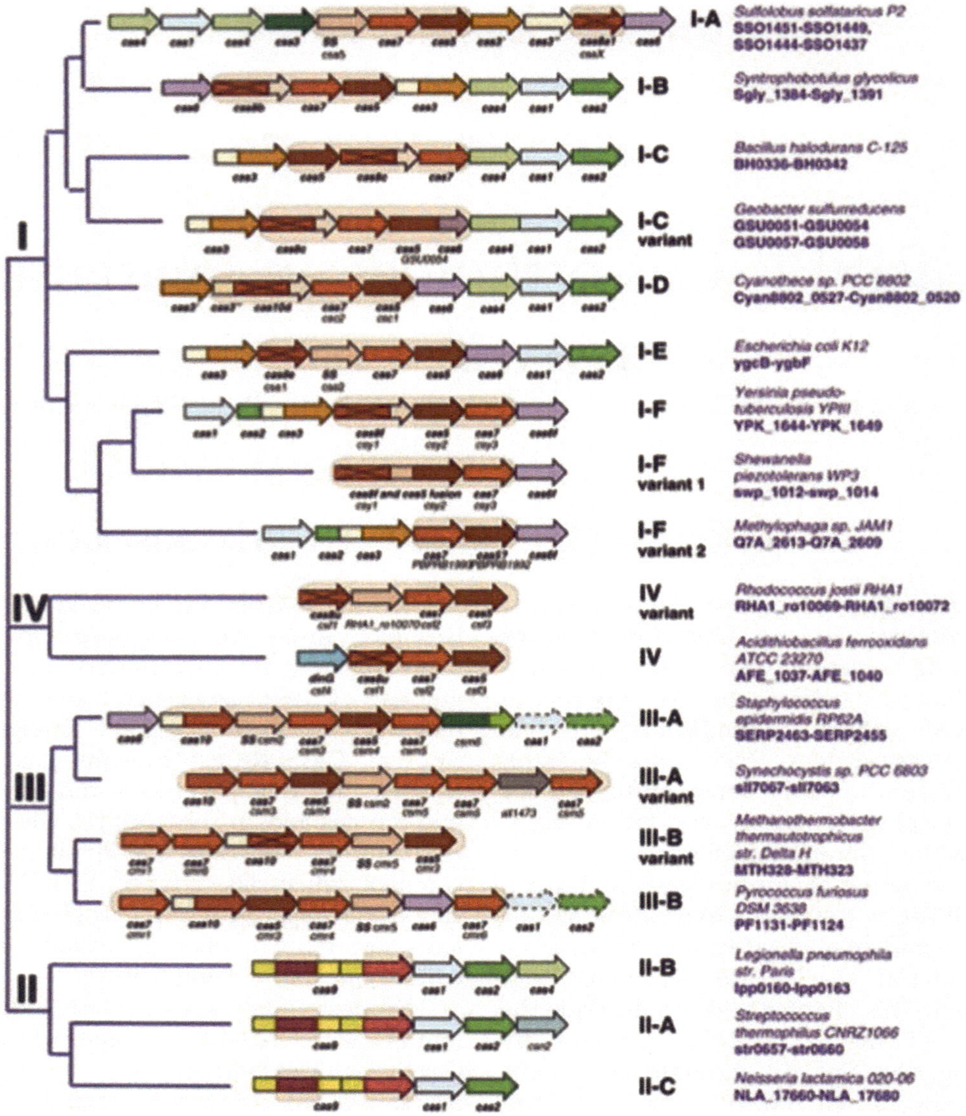

FIGURE 2.1 Classification and organization of CRISPR-Cas systems. Typical operon organization is shown for each CRISPR-Cas subtype. For each CRISPR-Cas subtype, a representative genome and the respective gene locus tag names are indicated. Homologous genes are color coded and identified by a family name. Names follow the classification from Ref. [3]. Names in bold are proposed systematic names; "legacy names" are in regular font.
LS, large subunit (including subfamilies of Cas10, Cas8, Cse1, Csy1); *SS*, small subunit (including Cmr2, Cmr5, Cse2).
Genes coding for inactivated large subunits are indicated by crosses. Genes and domain components for effector complexes are highlighted by pink background. *Source: Reproduced with permission from K.S. Makarova, E.V. Koonin, Annotation and classification of CRISPR-Cas systems. Methods. Mol. Biol. 1311 (2015) 47–75 [3]. The figure is reproduced under the terms set forth by the Springer Nature and Copyright Clearance Center: Rightslink, License Number 5112510578992, License Date July 19, 2021.*

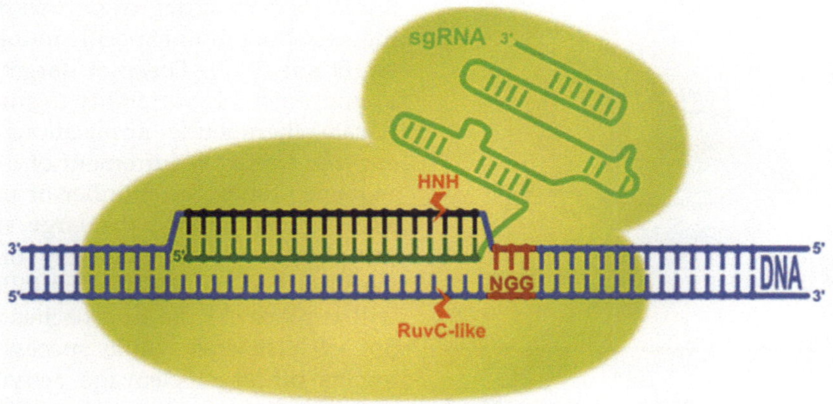

FIGURE 2.2 Genome editing through programmable RNA-guided DNA endonuclease cleavage with type II CRISPR/Cas9 system from *Streptococcus pyogenes*. The two-component system consists of Cas9 and of a single guide RNA (sgRNA) molecule, which was engineered from the dual tracrRNA:crRNA. sgRNA retains both critical features: the 20-nucleotide (nt)-long specific sequence at the 5′ end (*dark green*) and stem loop RNA structure at the 3′ end (*light green*), which is needed for binding to the Cas9 enzyme. On assembly of sgRNA and Cas9, the complex first recognizes the dinucleotide protospacer adjacent motif sequence (NGG, PAM, *red*) of DNA and unwinds it to enable Watson–Crick RNA–DNA pairing of complementary bases between target DNA, the so-called protospacer (*dark blue*) and the 20-nt guide sequence of sgRNA (*dark green*). Double-strand cleavage is performed by two Cas9 cleavage domains: the HNH domain cleaves the DNA strand that is complementary to the guide sequence of the sgRNA and the RuvC-like domain cleaves the DNA strand opposite the complementary strand, both cleaves three bases upstream (*orange arrows*) of the PAM sequence. *Source: Reproduced with permission from J. Murovec, Z. Pirc, B. Yang, New variants of CRISPR RNA-guided genome editing enzymes, Plant Biotechnol. J. 15 (8) (2017) 918 [10]. This figure is reproduced under the terms of the Creative Commons Attribution 4.0 International License (http://creativecommons.org/licenses/by/4.0/).*

(sgRNA/gRNA), which can be used to direct Cas9 to target loci within the genome [4,6–9]. The sgRNA is a synthetic sequence of approximately 20 nucleotides (nt) that defines the specificity of Cas9 by binding to target DNA via complementary base pairing [4]. Thus, the CRISPR-Cas system demonstrates a great deal of flexibility from a gene-editing perspective, given the possibility to change genome targets simply by altering sgRNA sequences. An additional requirement of the system is a protospacer adjacent motif (PAM), a short DNA sequence that enhances the specificity of the Cas protein (Fig. 2.2).

When Cas9 generates a DSB, it produces blunt ends of DNA that are subsequently acted upon by the cell's endogenous repair machinery. The most active repair mechanism is nonhomologous end joining (NHEJ), wherein break ends are directly ligated without a homologous template. NHEJ is an efficient but error-prone process that can introduce indel mutations at the target site [11]. Similarly, microhomology-mediated end joining, a process that uses 5–25 bp microhomologous sequences to align and repair broken ends, can also result in indels [12,13]. These repair mechanisms can lead to a diverse array of deletion, insertion, and frameshift mutations, resulting in potentially deleterious loss- or gain-of-function effects. Alternatively, if a homologous DNA template is present, the cell can implement specific alteration to the target sequence through homology-directed repair (HDR) mechanisms [14]. In HDR, single-stranded oligodeoxynucleotides (ssODNs) or double-stranded DNA constructs with homologous sequence flanking the target site can be used as repair templates [8,15] (Fig. 2.3).

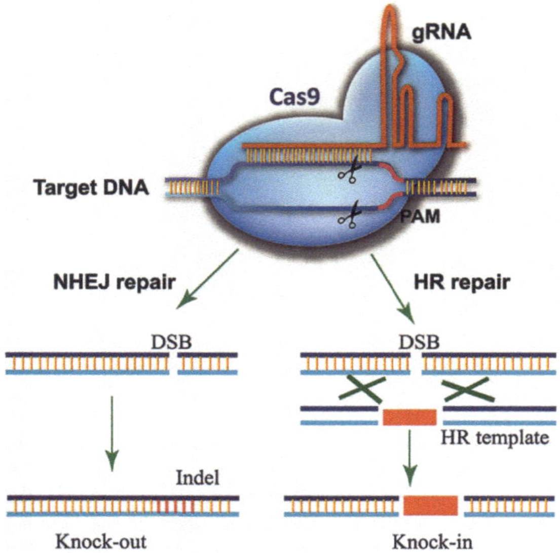

FIGURE 2.3 Cas9 is directed to its DNA target by base pairing between the guide RNA (gRNA) and DNA. A protospacer adjacent motif (PAM) downstream of the gRNA-binding region is required for Cas9 recognition and cleavage. Cas9/gRNA cuts both strands of the target DNA, triggering endogenous double-stranded break (DSB) repair. For a knock-out experiment, the DSB is repaired via the error-prone nonhomologous end joining (NHEJ) pathway, which introduces an indel at the DSB site that knocks out gene function. In a knock-in experiment, the DSB is repaired by homology-directed repair (HDR) using the donor template present, resulting in the donor DNA sequence integrating into the DSB site. *Source: Reproduced with permission from Ding et al., Recent advances in genome editing using CRISPR/Cas9, Front. Plant Sci. 7 (2016) 703 [16]. This figure is reproduced under the terms of the Creative Commons Attribution 4.0 International License (http://creativecommons.org/licenses/by/4.0/).*

Among the variety of species with CRISPR-Cas machinery, the Class 2 type II system of *S. pyogenes* was one of the first of its kind characterized in bacteria [17] and optimized for genome editing in human cells [6–9]. It consists of a Cas9 protein approximately 1370–1420 amino acids (aa) in length and requires a PAM sequence of NGG [6,8,9]. SpCas9 has been utilized to edit a wide range of human and animal models [18]; however, it has also had various shortcomings. First, it can tolerate a certain degree of base pair mismatching, resulting in nonspecific mutagenesis at off-target sites [13]. These off-target effects result in unpredictable variability in studies and complicate therapeutic applications. Second, the enzyme's strict requirement of a specific PAM sequence limits the number of potential target sites [17,19]. Finally, the large size of SpCas9 complicates efficient delivery of the complex via viral vectors [20,21]. The following section will explore (1) the approaches that maintain SpCas9 efficiency while successfully decreasing its off-target cleavage activity (Table 2.1) and (2) related Cas proteins with advantageous properties that address the challenges described above while opening up new avenues for gene-editing technology and application (Table 2.2).

Nickase SpCas9

Initial efforts to improve SpCas9 specificity resulted in SpCas9 nickase (nSpcas9), which cleaves a single strand of DNA due to silencing mutations in the catalytic residues of Cas9 (HNH [D10A] or RuvC [H840A]) (Fig. 2.4) [8,22,23]. By targeting paired nickases to opposite, proximal strands of DNA, a DSB is produced that can be repaired by either NHEJ or HDR in the cell [25,67]. The benefit of this approach is that repair of single-stranded nicks in the genome is more likely achieved by HDR while DSB repair is often dependent on the more error-prone NHEJ pathway [68]. Furthermore, this approach intrinsically reduces the likelihood of off-target effects because it is unlikely that two off-target sites will be generated close enough to one another to produce a DSB. The system is dependent on the design of highly active, target-specific sgRNAs that further decrease the possibility of off-target mutations [25]. Use of traditional repair templates of >200 nt ssDNA oligos is not recommended because compared to conventionally produced

TABLE 2.1 Characteristics of *Streptococcus pyogenes* synthetic Cas9 variants.

CRISPR-Cas orthologs or variants	Abbreviation	Size (aa)	Protospacer adjacent motif	Target length (nt)	First published
Streptococcus pyogenes Cas9	SpCas9	1368–1424	NGG	19–20	Jinek et al. [4]; Cong et al. [8]
S. pyogenes Cas9 nickase	nSpCas9	1368–1424	NGG	20	Cong et al. [8]; Mali et al. [22]; Ran et al. [23]; Cho et al. [25]
S. pyogenes Cas9 variant VQR	SpCas9-VQR	1372	NGAN, NGNG	19–20	Kleinstiver et al. [19]
S. pyogenes Cas9 variant EQR	SpCas9-EQR	1372	NGAG	19–20	Kleinstiver et al. [19]
S. pyogenes Cas9 variant VRER	SpCas9-VRER	1372	NGCG	20	Kleinstiver et al. [19]
S. pyogenes Cas9 variant QQR1	SpCas9-QQR1	1372	NAAG	20	Anders et al. [26]
S. pyogenes Cas9 variant xCass93.7	xCas9	1368	NG, NNG, CAA, GAT, GAA	20	Hu et al. [27]
S. pyogenes Cas9 variant NG	SpCas9-NG	1368	NGA, NGT, NG	20	Nishimasu et al. [28]
S. pyogenes Cas9 high fidelity (HF)	SpCas9-HF1	1368	NGG	20	Kleinstiver et al. [29]
S. pyogenes Cas9 enhanced-specificity	eSpCas9	1424	NGG	20	Slaymaker et al. [31]
S. pyogenes deactivated-Cas9	dCas9	1368–1424	NGG	20	Bickard et al. [32]; Qi et al. [33]
S. pyogenes deactivated-Cas9-*FokI*	dCas9-*FokI*	1817	NGG	20	Tsai et al. [34]; Guilinger et al. [35]
S. pyogenes deactivated-Cas9-Tet1 and -Dnmt3a	dCas9-Tet1, dCas9-Dnmt3a	2115, 2309	NGG	20	Liu et al. (2016) [36]
S. pyogenes Cas9 cytidine deaminases	BE1, BE2, BE3, BE4	1623–1710	NGG	20	Komor et al. [37]; Nirshida et al. [38]; Kim et al. [39]
S. pyogenes Cas9 adenine base editors	ABE7.9, 7.10	1775	NGG	20	Gaudelli et al. [41]

Modified from J. Murovec, Z. Pirc, B. Yang, New variants of CRISPR RNA-guided genome editing enzymes, Plant Biotechnol. J. 15 (8) (2017) 918 [10]; A. Cebrian-Serrano, B. Davies, CRISPR-Cas orthologs and variants: optimizing the repertoire, specificity and delivery of genome engineering tools, Mamm. Genome 28 (7–8) (2017) 253 [42]; D.J.W. Brocken, M. Tark-Dame, R.T. Dame, dCas9: a versatile tool for epigenome editing, Curr. Issues Mol. Biol. 26 (2018) 15–32 [43]; D.J.W. Brocken, M. Tark-Dame, R.T. Dame, dCas9: A versatile tool for epigenome editing, Curr. Issues Mol. Biol. (2018) 15–32; K. Liu, C. Petree, T. Requena, P. Varshney, G.K. Varshney, Expanding the CRISPR toolbox in Zebrafish for studying development and disease, Front. Cell Dev. Biol. 7 (2019) [44].

TABLE 2.2 Overview of Cas orthologs and synthetically redesigned endonucleases.

CRISPR-Cas orthologs or variants	Abbreviation	Class and type	Size (aa)	Protospacer adjacent motif	Target length (nt)	First published
Staphylococcus aureus Cas9	SaCas8	Class 2 type II	1053	NNGRRT	20–24	Ran et al. [45]
S. aureus Cas9 vairant KKH	SaCas9-KKH		1053	NNNRRT	21	Kleinstiver et al. [46]
S. aureus Cas9 nickase	nSaCas9		1053	NNGRRT	24	Friedland et al. (2015) [47]
Streptococcus thermophilus 1 Cas9	St1Cas9	Class 2 type II	1122	NNAGAAW	20	Muller et al. [48]
S. thermophilus 3 Cas9	St3Cas9	Class 2 type II	1393	NGGNG	20	Glemzaite et al. [49]
Neisseria meningitidis Cas9	NmCas9 (or NmeCas9)	Class 2 type II	1109	NNNNGMTT	23–24	Hou et al [50]; Fonfara et al. [51]
Francisella novicida Cas9	FnCas9	Class 2 type II	1629	NGG	22	Fonfara et al. [51]; Hirano et al. [53]
F. novicida Cas9 variant RHA	FnCas9-RHA		1632	YG	22	Hirano et al. [53]
Campylobacter jejuni Cas9	CjCas9	Class 2 type II	984	NNNVRYAC, NNNNRYAC, NNNVRYM	22	Kim et al. [39,40]; Yamada et al. [54]
Treponema denticola Cas9	TdCas9	Class 2 type II	1423	NAAAAN	20	Esvelt et al. [55]
Streptococcus macacae Cas9	SmacCas9	Class 2 type II	1338	NAAN	20	Chatterjee et al. (2018) [56]
Streptococcus canis Cas9	ScCas9	Class 2 type II	1368	NNG	20	Chatterjee et al. (2018) [56]
S. canis Cas9 variant LoopKQ	ScCas9-KQ		1375	NNGA, NGG	20	Chatterjee et al. (2018) [56]
Streptococcus mutans Cas9	SmCas9	Class 2 type II	1345	NNG	20	Gong et al. (2018) [57]
Acidaminococcus Cas12a	AsCas12a/Cpf1	Class 2 type V	1308–1310	TTTV	23–24	Zetsche et al. [58]
Lachnospiraceae Cas12a	LbCas12a/Cpf1	Class 2 type V	1228	TTTV	23–24	Zetsche et al. [58]

Francisella Cas12a	FnCas12a/Cpf1	Class 2 type V	1159	TTN, KYTC	23–24	Zetsche et al. [58]; Tu et al. [59]
Moraxella Cas12a	MbCas12a/Cpf1	Class 2 type V	1300	TTN	23–24	Zetsche et al. [60]
AsCas12a, LbCas12a, FnCas12a, and MbCas12a variant RR				TYCV, TWTV	23–24	Gao et al. [62]; Nishimasu et al. [63]; Toth et al. (2018)
AsCas12a, LbCas12a, FnCas12a, and MbCas12a variant RVR				TATV	23–24	Gao et al. [62]; Nishimasu et al. [63]; Toth et al. (2018) [64]
Leptotrichia buccalis Cas13a	LbCas13a/C2c2	Class 2 type VI	1159	3′ A, U, or C (PFS)	24	East-Seletsky et al. (2016) [65]
Leptotrichia shahii Cas13a	LsCas13a/C2c2	Class 2 type VI	1389	3′ A, U, or C (PFS)	24	Abudayyek et al. [66]

Notes: D = A, G, or T; N = any base; R = G or A; W = A or T; Y = C or T. *PFS*, protospacer-flanking sequence, analogous to PAM for Cas9.

Modified from J. Murovec, Z. Pirc, B. Yang, New variants of CRISPR RNA-guided genome editing enzymes, Plant Biotechnol. J. 15 (8) (2017) 918 [10]; A. Cebrian-Serrano, B. Davies, CRISPR-Cas orthologs and variants: optimizing the repertoire, specificity and delivery of genome engineering tools, Mamm. Genome 28 (7–8) (2017) 253 [42]; D.J.W. Brocken, M. Tark-Dame, R.T. Dame, dCas9: a versatile tool for epigenome editing, Curr. Issues Mol. Biol. 26 (2018) 15–32 [43]; D.J.W. Brocken, M. Tark-Dame, R.T. Dame, dCas9: A versatile tool for epigenome editing, Curr. Issues Mol. Biol. (2018) 15–32; K. Liu, C. Petree, T. Requena, P. Varshney, G.K. Varshney, Expanding the CRISPR toolbox in Zebrafish for studying development and disease, Front. Cell Dev. Biol. 7 (2019) [44].

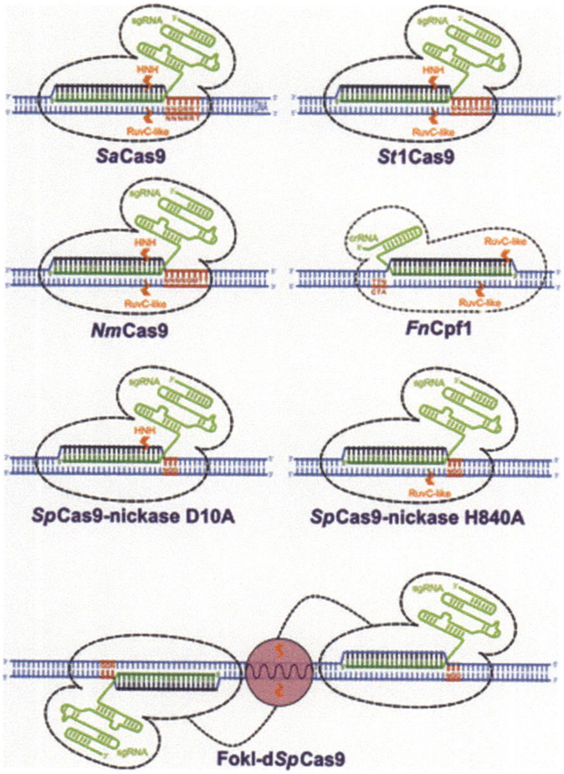

FIGURE 2.4 Different variants of DNA-modifying enzymes derived from CRISPR-Cas and CRISPR-Cpf1. Enzymes bind to DNA based on recognition of protospacer adjacent motifs (*red*) and Watson–Crick RNA–DNA pairing of complementary bases between DNA protospacers (*dark blue*) and sgRNA or crRNA guide sequences (*dark green*). Cleavage of single or both strands of DNA is performed by RuvC-like, HNH or *FokI* nuclease domains (*orange*). *Source: Reproduced with permission from J. Murovec, Z. Pirc, B. Yang, New variants of CRISPR RNA-guided genome editing enzymes, Plant Biotechnol. J. 15 (8) (2017) 920 [10]. This figure is reproduced under the terms of the Creative Commons Attribution 4.0 International License (http://creativecommons.org/licenses/by/4.0/).*

DSBs, longer distances between nick sites reduce the effective length of the homology arms [69]. However, Ge and Hunter demonstrate that use of long single-strand DNA donor templates addresses the issue of extended spacing between nick sites by allowing for longer insertions [69]. The paired nickase method has successfully edited mammalian cell lines and mouse zygotes without inducing any detectable, undesirable off-target effects [8,24,25,70,71] and has shown promise as an efficient way to model and treat disease. Biallelic modification of a locus in a patient-derived induced pluripotent stem cells (iPSCs) has been reported [67], including successful integration of a GFP reporter in the locus of PAX7 [72]. Wu et al. produced a guideline outlining the nickase approach to generate and validate knock-ins in iPSCs [72]. Phenotypes associated with a deficiency in FAM83H [24] and the knock-out of schizophrenia-related locus Fabp7 [70] were successfully produced using the paired nCas9 approach. Additionally, nickase SpCas9 mediated gene correction in fibroblasts derived from a patient with Fanconi anemia (FA), and did so with a higher frequency than wild-type (WT) SpCas9 [73]. Van de Vrugt et al. expanded the study of FA by demonstrating that although gene-editing efficiency using dual nickase systems was low, FA-corrected embryonic stem cells (ESCs) acquired a strong proliferative advantage over nonedited cells [74]. Paired nCas9s demonstrated a recombination efficiency of 66.7% when used to correct a F9 gene deficiency in a mESC model of Hemophilia B, with all of the corrected target sites demonstrating site-specific integration of the WT locus [75]. Insertion of polyadenylation signals via paired nCas9 upstream of deleterious CTG repeats in a model of myotonic dystrophy type 1 led to phenotype reversal in differentiated human neural and muscle cells, suggesting that this may be a viable therapeutic strategy [76]. In a model of hepatitis B virus (HBV), Cas9 nickase vector systems targeting HBV loci significantly reduced viral replicative intermediates and extracellular HBV surface and envelope antigens while producing no detectable off-target mutations [77]. Targeted paired-nickase cleavage of a neuroblastoma-associated locus in a mouse Neuro2A cell line resulted in a significant decrease in cell proliferation [78], and knock-out of metastatic-promoting factors in ovarian cancer cell lines exhibited specific biomarkers that are promising for future clinical therapies [79].

Comparisons of efficacy and specificity between WT SpCas9 and paired nickases have had conflicting results. Bialk et al. found that compared to nickases, WT nucleases exhibited a higher level of gene-editing activity at a mutant eGFP locus in human colon cancer cells. However, this result has not been confirmed across different loci and cell types [80]. As HDR is central to precisely editing the genome, SpCas9 and paired nickase platforms were evaluated to determine which conditions favored HDR over NHEJ. In general, wtSpCas9 had higher activity, especially when initiating NHEJ, than Cas9 nickases; however, genome-editing outcomes were predominantly condition dependent [81]. For example, in HEK293 cells, Cas9 nickases induced more HDR than NHEJ, but HDR was undetectable in HeLa cells [81]. This finding questions the validity of the paired nickase approach as an effective strategy for achieving high gene-editing specificity.

High-fidelity and enhanced SpCas9

To further improve specificity and eliminate off-target recognition, the DNA-binding domain of SpCas9 was redesigned to reduce nonspecific DNA interaction. Structural studies found that the SpCas9–sgRNA–target DNA complex includes direct hydrogen bonds between four SpCas9 residues (N497, R661, Q695, and Q926) and the phosphate backbone of the target DNA (Fig. 2.5) [82,83]. Kleinstiver et al. hypothesized that disruption of these contacts might alter the SpCAs9–sgRNA complex so as to retain enough on-target activity while diminishing off-target mismatches. Variants were constructed with selective point mutations in DNA-binding regions and on- and off-target cleavage was characterized [29]. One construct (SpCas9-HF1, N497A/R661A/Q695A/Q926A) demonstrated 70% of the activity of wtSpCas9 but substantially higher genome-wide specificity with nearly undetectable levels of off-target mutagenesis [29]. Compared to wtSpCas9, SpCas9-HF1 increased HDR events while reducing indel formation when used to introduce monoallelic and biallelic point mutations into a mammalian cell line [84]. Additionally, Slaymaker et al. suggested that neutralizing the positive charges between the cleavage and PAM-interacting domains of SpCas9 could reduce the energy of the nontarget strand binding [31]. They characterized two variants [SpCas9-K855A, eSpCas9 (K848A/K1003A/R1060A)] that when compared to wtSpCas9, have comparable efficacy but reduced off-target cleavage effects. Later studies found that SpCas9-HF1 and eSpCas9 are trapped in an inactive state when bound to mismatched targets, indicating that these high-fidelity Cas9 variants harbor amino acid substitutions that reduce initiation of cleavage at sites that do not exactly complement the sgRNA sequence [85]. It was found that a noncatalytic Cas9 domain, REC3, assesses target complementarity and regulates the HNH nuclease domain. Chen et al. proposed that high-fidelity variants of Cas9 reduce off-target cleavage by raising the threshold for HNH conformation activation when bound to substrates [85].

The initial success of SpCas9-HF1 and eSpCas9 motivated further engineering and experimentation. To address a subset of targets at which SpCas9-Hf1 and eSpCas9 were still susceptible to producing off-target effects, Kulcsar et al. designed "highly enhanced fidelity" Cas9 (HeFSpCas9), which contains both eSpCas9 and SpCas9-HF1 mutations. Another high-fidelity variant, HypaCas9, demonstrated better genome-wide specificity than wtSpCas9, SpCas9-HF1, and eSpCas9 [85]. Casini et al. produced "evolved" Cas9, evoCas9 (K526E) and compared it to the wtSpCas9 and other high-fidelity variants in side-by-side testing. Overall, evoCas9 eliminated 98% of detectable off-targets associated with standard and

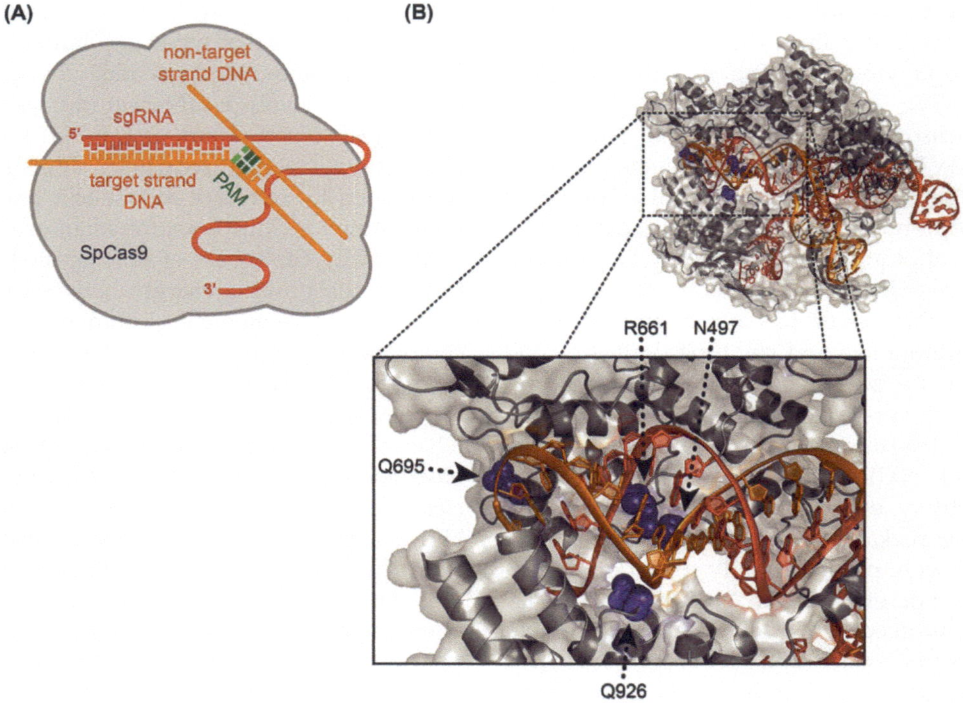

FIGURE 2.5 (A) Schematic illustrating the SpCas9–sgRNA complex, with base pairing between the single guide RNA (sgRNA) and target DNA. (B) Structural representation of the SpCas9–sgRNA complex bound to the target DNA. The four residues that form hydrogen bond contacts to the target-strand DNA backbone are highlighted in blue; the HNH domain is hidden for visualization purposes. *PAM*, Protospacer adjacent motif. *Source: Reproduced with permission from B.P. Kleinstiver, V. Pattanayak, M.S. Prew, S.Q. Tsai, N. Nguyen, Z. Zheng, et al., High-fidelity CRISPR-Cas9 variants with undetectable genome-wide off-targets, Nature 529 (7587) (2016), 490–495 [29]. The figure is reproduced under the terms set forth by the Springer Nature and Copyright Clearance Center: Rightslink, License Number 4644310647954, License Date August 8, 2019.*

clinically relevant genomic loci while maintaining near WT levels of activity [86]. In all but three loci, evoCas9 was more active than SpCas9-HF1 [86]. In a metaanalysis, evoCas9 was found to generate less off-target cleavage events than HypaCas9 at three loci (12 vs 20 total off-target loci for evoCas9 and HypaCas9, respectively) and showed a better on-/off-target ratio for 7 out of 9 commonly detected off-targets [86]. Another high-fidelity variant is HiFiCas9 (R691A), which demonstrated a strong adeno-associated virus (AAV)6-mediated gene-targeting efficiency at five clinically relevant loci (HBB, IL2RG, CCR5, HEXB, and TRAC) in human cell lines [87]. Compared to wtSpCas9, eSPCas9, SpCas9-HF1, and HypaCas9, HifiCas9 showed improved on-target editing at sites when delivered as a ribonucleoprotein complex [87]. When used to target Glu6Val mutations associated with sickle cell anemia, HifiCas9 corrected almost as many mutations as WT (64% vs 70%) and reduced off-target cutting by 20-fold [87].

Despite the proliferation of high-fidelity variants, there are limitations to these systems that must be considered. Both eSpCas9 and SpCas9-HF1 have stricter requirements for sgRNA guide sequences than WTSpCas9, that is, perfectly matched 20-nt sgRNAs are necessary to enable robust gene editing [88]. When additional matched or mismatched residues are added to the 5′ end of the gRNA, SpCas9-HF1 and eSpCas9 activity decreases dramatically [88]. Additionally, Kulcsár et al. demonstrate that for the highest specificity cleavage, each target must be matched with an appropriately selected nuclease [89]. They propose a framework for generating new high-fidelity variants for targets that have no optimal nuclease—this entails pretesting the candidate target with various high-fidelity nucleases to determine which exhibits adequate activity for the desired application [89].

SpCas9 variants with novel PAM requirements

The specificity of SpCas9 is further determined by the nature of its PAM sequence, which, in this case, is a sequence of NGG that lies directly downstream (3′) of the target site. The PAM sequence is considered a strict requirement of the CRISPR-Cas system but only occurs on average once in every 16 randomly chosen genomic loci, thus limiting the number of targets that can be established. This presents a particular challenge when targeting A/T-rich regions of the genome and for applications that require precise editing parameters like base editing and homology-directed repairs, both of which have very specific PAM requirements. Base editing is optimal when the PAM sequence is approximately 13–17 nt from the target [41,90] while homology-directed repairs seem to be most efficient when cleavage occurs approximately 10–20 nt from a chosen target [91,92]. It has been reported that wtSpCas9 can, to an extent, recognize alternative PAMs, 3′ NAG and NGA [19,93]. However, these noncanonical recognition events are not programmable and thus could result in unwanted off-target effects. Kleinstiver et al. found that a D1135E mutation in the PAM recognition residue decreases SpCas9 NAG and NGA recognition, improving canonical PAM recognition and decreasing the likelihood off-target mutagenesis [19].

However, engineered variants with predictable behavior are desirable for increasing target possibilities while maintaining SpCas9 specificity. SpCas9 mutants have been engineered to recognize alternative PAM sequences. SpCas9 variants with comparable specificity to wtSpCas9 have been engineered to recognize NGCG (variant VRER, D1135V/G1218R/R1335E/T1337R), NGAG (variant VQR, D1135V/R1335Q/T1337R), and NGAG (variant EQR, D1135E/R1335Q/T1337R) [19]. A comparison of VRER, VQR, EQR, and wtSpCas9 revealed that in the variants, the multiple mutations displace the phophodiester backbone of the PAM duplex and allow for direct recognition of the altered PAM nucleotides [52]. The crystal structures of the three variants demonstrate an overall similarity to wtSpCas9, suggesting that the mutations do not substantially affect the architecture of the Cas9–sgRNA–PAM complex [52]. An engineered SpCas9 variant QQR1 demonstrates a preference for NAAG PAMs but also displays slower cleavage activity than that of the WT [26]. More recently, an expanded PAM SpCas9 variant, xCas9, was developed, and it recognizes a broader range of sequences including NGX, GAA, and GAT [27]. xCas9 showed comparable activity to wtSpCas9 when tested on NGG PAM sites and about a 2-fold increase in activity when tested on NGT, NGA, and NGC PAMs. For GAA and GAT sites, wtSpCas9 had nearly undetectable levels of activity while xCas9 showed at least a low level of indel production. Genome-wide unbiased identifications of DSBs evaluated by sequencing (GUIDE-seq), a tool that relies on the integration of double-stranded oligodeoxynucleotides into DSBs to map off-target effects, found that xCas9 results in much lower off-target activity than SpCas9. The authors found this surprising given that xCas9's broadened PAM compatibility should increase promiscuity [27]. They suggest that the DNA promiscuity of wtSpCas9 probably evolved to better fight off viral invasion, but these characteristics can be overcome with engineering. Although a promising avenue for target modification, the efficacy of xCas9 on noncanonical PAMs varies based on the application [27].

Deactivated-Cas9 characterization and application

Previously discussed variants leveraged properties of SpCas9 to improve specificity while maintaining cleavage activity. Alternatively, a deactivated-Cas9 (dCas9), with mutated catalytic residues, has demonstrated its usefulness in designing precise and programmable DNA-binding applications. Originally engineered to elucidate the catalytic domains of Cas9 responsible for cleavage [4], dCas9 has been utilized for transcriptional blockage and activation, gene modulation, epigenetic editing, and cell staining.

Different dCas9 strategies have been developed to manipulate the genome and epigenome in an RNA-dependent, site-specific manner. In the simplest approach, sgRNA–dCas9 complexes are used to target and block gene transcription via steric hindrance [32,33,94,95]. In prokaryotes, this CRISPR-interference strategy demonstrated an approximate 300-fold decrease in mRNA production when a single sgRNA was used to guide dCas9 and an approximate 1000-fold decrease when two sgRNAs were used in combination [33]. However, this direct transcriptional interference was not successful in mammalian cell lines [33], prompting development of a more sophisticated gene-regulation approach in which recombinant effector proteins are generated via fusion of transcription factor domains to dCas9. Fusion of transcription activator domains (VP64, p65, Rta) or transcription repressor domains (KRAB, SID) successfully modulates gene expression [96–100]. Moreover, fusion of the three transcription activator domains in succession to dCas9 (VPR-dCas0) maximizes transactivation effects to an almost 100-fold increase compared to dCas9-VP64 alone [101] (Fig. 2.6). VPR-dCas9 was used in melanoma and breast cancer cell lines to induce expression of the PTEN tumor suppressor gene, which repressed downstream oncogenic signaling [102]. This "direct effector" approach has also recently been optimized for up-regulation of target genes in neuronal cells in vitro and in vivo [103]. In addition to transcription factors, epigenetic effectors have also been fused to dCas9. Methyltransferase 3A (DMT3A) [104,105], TET1 5mC-hydroxylase [97,106], LSD1 histone demethylase [107,108], and p300 and CBP histone acetyltransferases [109,110] have all been successfully used as epigenetic editing tools.

An indirect effector approach entails fusing additional motifs that recruit effector proteins to dCas9. One example is SunTag, a protein scaffold capable of recruiting effector domains via specific single-chain variable fragment antibodies. SunTag that is fused to Vp64 can successfully recruit dCas9 and produce more gene expression compared to dCas9-Vp64 alone [111]. Another example is the use of scaffolding RNA (scRNA), which is a complex created by fusing sgRNAs to RNA aptamers, secondary RNA units that are targeted by RNA-binding proteins [22,106,109,112,113]. By binding to scRNA, transcription effectors are recruited to the dCas9 [22,109,113]. Combining the two strategies, dCas9–effector fusion and apatamer-recruitment, further amplifies downstream transcriptional events [106,112]. Konermann et al. recruited two RNA-binding transcription factors to dCas9-VP64 via scRNA, producing a 100-fold intensification of transcriptional activation compared to dCas9-VP64 alone [112]. The study also found that one scRNA performed better than eight sgRNAs when targeting dCas9-VP64 to promoter

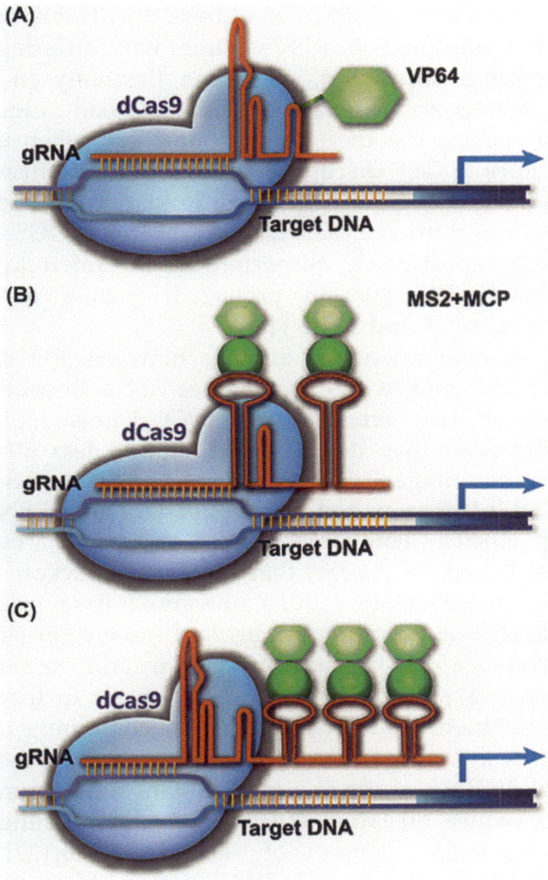

FIGURE 2.6 (A) The dCas9-activation domain (AD) is guided to the promoter of a target gene, activating transcription. (B, C) A three-component system for target gene activation. The gRNA is modified by adding an RNA ligand, MS2, to its loop or 3′-end, without impairing its binding to dCas9. MS2 bacteriophage coat protein (MCP), which specifically binds the MS2 ligand, is fused with the AD fragment. The dCas9/gRNA-MS2 is directed to the promoter of the target gene and recruits MCP-AD to activate transcription. *Source: Reproduced with permission from Y. Ding, H. Li, L.-L. Chen, K. Xie, Recent advances in genome editing using CRISPR/Cas9, Front. Plant Sci. 7 (2016) 703 [16]. This figure is reproduced under the terms of the Creative Commons Attribution 4.0 International License (http://creativecommons.org/licenses/by/4.0/).*

regions, emphasizing the strength of this approach [112].

Deactivated-Cas9–effector complexes can also be utilized for spatiotemporal control of activity. One strategy is inducible expression of dCas9 or the sgRNA with doxycycline, although this approach results in a somewhat leaky expression due to absence of an inducer [114–116]. For more precise temporal control, split-dCas9 or split-dCas9–effector proteins can be conditionally assembled into functional complexes through different induction strategies. In a truncated-sgRNA strategy, split sgRNAs prevent the joining of a split-dCas9 domain or stimulate complex disassembly, while the presence of a full-length sgRNA generates a functional complex from two split-dCas9 domains [117]. Chemical induction of gene expression can be achieved by fusing dCas9 to rapamycin-binding domains [118]. Finally, construction of a dCas9–effector complex with photoinducible dimerization domains results in light activation of the system [119].

Although studies on target specificity have predominantly focused on wtSpCas9, these same findings can be extended to dCas9, as it only differs from WT at its mutated catalytic residues [30,31,120]. Indeed, dCas9 exhibits off-target binding on a genome-wide scale [121,122]. Adopting engineering strategies to modify wtSpCas9 to ensure higher fidelity or to create novel PAM recognition could address the limitations of dCas9 specificity. Another concern of dCas9 is that it is toxic in many bacteria, thus limiting the levels at which it can be expressed in experimental models [123]. A nontoxic version of dCas9 has been engineered which allowed approximately 17-fold higher levels of dCas9 to be expressed in *Escherichia coli* compared to conventional dCas9 [123]. Finally, the coding sequence of the commonly used dSpCas9 is around 4.2 kb; although this sequence size is compatible with the ~4.5 kb maximum accommodation of viral vectors used for delivery in mammalian cells, it is not compatible for the codelivery of sgRNA sequences and fused-effectors [43]. The challenge of size can be addressed by delivering split Cas domains [118] or smaller Cas orthologs, which will be addressed later in the chapter.

FokI-*dCas9: a dCas9 application to improve specificity*

Another approach to achieve higher editing specificity with CRISPR systems is to use RNA-guided dCas9-*Fok*I nucleases (RFNs). Gene-editing approaches like zinc-finger nucleases and transcription activator-like effector nucleases rely on dimeric activation to cleave DNA. The requirement of two separate monomers to simultaneously bind and produce one cleavage event intrinsically lessens the likelihood of off-target events (Fig. 2.4). This concept was applied to CRISPR technology and resulted in the production of the RFN system. Monomeric *FokI* nuclease domains are not catalytically active [124], and through fusion to a dCas9, an obligate dimeric RFN is produced. DNA can only be cleaved when two distinct RFNs are bound to adjacent sites within a specified sgRNA distance and orientation ("PAM-out" or N termini of dCas9s toward each other) [35,122]. The sgRNA spatial distance, or the number of base pairs between the two sgRNA-binding sites, is relatively stringent for RFNs, falling between 14 and 29 base pairs [34,35,125]. Overall, as summarized by Havlicek et al., this means that a target sequence must fit the motif of $CCNN_{20}$ (*left*, nontarget strand), 14–29 bp random spacer sequence, and $N_{20}NGG$ (*right*, target strand), which is a requirement that few genomic loci satisfy [125].

A key component of this set-up is the choice of the linker peptide between the *Fok*I and dCas9 domains. A 5 aa linker (GGGGS) limited activity of the RFN to only a few sgRNA spacing distances, 14–17 bp and 26 bp [34]. Another study testing 17 different fusion linkers over a set number of sgRNA distances resulted in most of the variants showing unexplainable inactivity or very weak activity, prompting questions of fusion linker influence on RFN activity [35]. To address this, Havlicek et al. designed six RFN variants with differing length, composition, and linker flexibility and characterized their activity over a broad range of spacing distances. They found that when it comes to GFP disruption assays and modification of endogenous genes in human cells, one variant with a 25 aa, flexible linker ($GGGGS_5$-RFN) consistently outperformed the reference GGGGS-RFN within preferred spacing distances of 19 and 29 bp [125].

A comparison of activity between RFNs, wtCas9, and nCas9 reveals that RFNs are ultimately less efficient than WT Cas9s [35]. However, they demonstrate 140-fold less off-target mutagenesis than wtCas9 and 1.3- to 8.8-fold less than paired nickases [35]. Higher RFN specificity compared to nCas9 may be explained by the fact that monomeric nickases can in principle induce unwanted base pair substitutions at target sites and have a larger tolerance for varied orientation and target spacing [34,35,122]. Wyvekens et al. further fine-tuned the RFN approach by combining it with a "tru-sgRNA" strategy—shortening the guide RNA at the 5′ end by 2–3 nt has shown to reduce off-target effects while maintaining comparable levels of activity [126,127]. Application of this "tru-RFN" method in mammalian and human cancer cell lines showed no statistically significant difference between tru-RFN and conventional RFN efficiency but tru-RFN did exhibit an approximate 40% reduction in unwanted indels [127].

Base editing

Generating single nucleotide changes via targeted knock-in mutations, even with the benefits of CRISPR-Cas editing tools, can be quite challenging due to the very low success rate of HDR and required presence of a precise donor template. CRISPR-mediated base editing, or the conversion of G to A and C to T, is a promising method for inducing irreversible point mutations without generating DSBs

[37,41]. Several studies have demonstrated the ability to fuse nCas9 or dCas9 to a cytidine deaminase, such as apolipoprotein B editing complex 1 or activation-induced cytidine deaminase. Like any Cas9 protein, this complex is guided by a sgRNA but can convert cytidine to uridine, effecting a C to T (or G to A) substitution [37–39,85,128–135]. In the original study, cytidine deaminase base editors proved capable of modifying human and murine cell lines, correcting approximately 15%–75% of the total cellular DNA with less than 1% of indel formation [37]. Further study displayed that base editing generated deletions and point mutations in mammalian cells just as conventional Cas9 [38]. This strategy has proven effective in editing yeast, plant, and mammalian genomes [38,39,85,128–135]. As there are no natural enzymes that convert A to G or T to C, a complicated series of engineering ensued to produce an adenine base editor (ABE), ultimately resulting in a variant, ABE7.1, which converts approximately 50% of AT to GC in mammalian cells [41].

The challenges of base editing mirror those of conventional SpCas9 use, including limitations in specificity, efficiency, and the PAM requirement. A second-generation base editor (BE2) was quickly developed after initial experiments and contains a uracil glycosylase inhibitor domain to prevent the deleterious effects of an excision of uracil during repair [37]. However, BE2 only demonstrated a marginal increase in efficiency compared to BE [37]. A more efficient base editor, BE3, was subsequently developed by restoring a catalytic His residue responsible for nicking the nonedited strand to mimic newly synthesized DNA [37]. To further eliminate off-target effects, high-fidelity variants HF2-BE2 and HF-BE3 were developed [34,125]. However, it was discovered that delivery of the BE3 as a ribonucleoprotein complex is more important to target-editing specificity than the production of a high-fidelity enzyme [131]. Base-editing specificity ratios for on-target:off-target editing increased ~66-fold with RNP delivery of BE3 compared with plasmid delivery of BE3 [131]. This result is in agreement with other mammalian studies that emphasize that RNP delivery of Cas9 markedly improves target-editing specificity in vitro and in vivo [136–138]. Development of the BE4 editor showed increased base conversion efficiency by 50% compared to prior generations [90], and subsequent improvement of BE3 and BE4 editors (BE3Gam, BE4Gam) improved conversion by another 50% [139]. Another base editor, eA3A-BE3, has shown activity that is comparable to that of other base editors while deaminating in a more controlled manner according to a TCR > TCY > VCN hierarchy [140]. To broaden the target window of base editors, a base editor with comparable efficiency to BE3 was developed. Known as BE-PLUS, this base editor utilized the SunTag system, discussed previously [133]. Both cytidine deaminase and adenine deaminase base editors have been fused with alternative PAM recognition variants (VQR, VRER, KKH, xCas9) to broaden their PAM compatibility [27,39].

CRISPR-Cas orthologs

An efficient delivery mechanism is necessary for gene-editing tools to work in a therapeutic context. In recent years, AAV vectors have been one of the most popular tools for gene therapy delivery. However, the maximum AAV cargo size is about 4.7 kb, which limits the potential of the large SpCas9 (~4.2 kb) to be packaged and delivered with its sgRNA. To solve the problem of delivery as well as to further explore options to increase gene-editing specificity and efficiency, researchers turned to orthologous CRISPR-Cas systems (Fig. 2.7). Initially focusing on other Class 2 type II systems, Ran et al. screened over 600 Cas9 orthologs and discovered some species had Cas9s that were significantly shorter than that of *S. pyogenes* (~1000 aa vs ~1350 aa), potentially

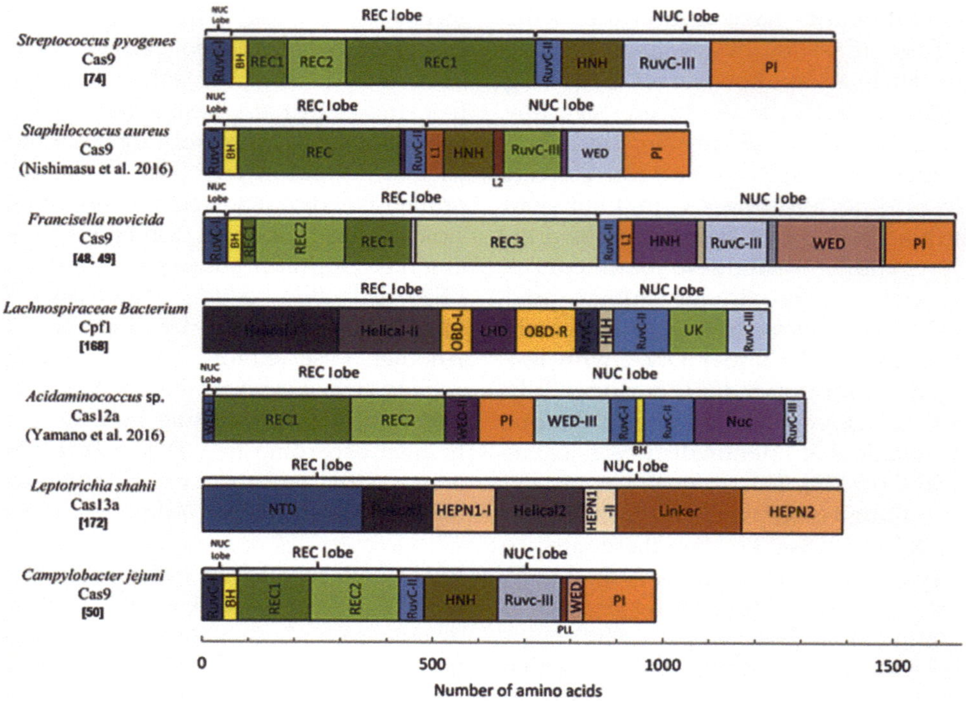

FIGURE 2.7 Domain structure of CRISPR-Cas orthologs. *BH*, Bridge helix domain; *LHD*, looped-out helical domain; *L1* and *L2*, linker 1 and 2; *NTD*, N-terminal domain; *NUC*, nuclease domain; *OBD*, oligonucleotide-binding domain; *PLL*, phosphate lock loop; *REC*, recognition domain; *UK*, unknown function domain; *WED*, wedge domain. *Source: Reproduced with permission from A. Cebrian-Serrano, B. Davies, CRISPR-Cas orthologs and variants: optimizing the repertoire, specificity and delivery of genome engineering tools, Mamm. Genome 28 (7–8) (2017) 253 [42]. This figure is reproduced under the terms of the Creative Commons Attribution 4.0 International License (http://creativecommons.org/licenses/by/4.0/).*

promising improved CRISPR-Cas delivery in AAV vectors [45]. Further studies have elucidated not only a number of promising Cas9 orthologs but also related effector nucleases from other Class 2 types (Table 2.2). As discussed below, the enzymes have differing sizes, PAMs, target lengths, and other specificities, broadening the overall potential pool of targets and applications.

CRISPR-Cas9 system orthologs

Staphylococcus aureus *Cas9*

S. aureus Cas9 (SaCas9) targets sequences between 21 and 24 bp and has a PAM requirement of NNGRRT (Fig. 2.4). The shorter size of SaCas9 (1053 aa) compared to that of SpCas9 makes it more compatible for packaging and delivery via an AAV vector. [45,141]. SaCas9 delivered by an AAV vector has generated mutant mice with successful gene disruption and knock-in modifications [45,141–147]. In human cell studies, SaCas9 has demonstrated efficient gene editing when used to knock-out disease-associated loci in hematopoietic stem and progenitor cells [148] and in human prostate cancer and motor neuron cells [149]. Initial studies found that in human and mouse cells, SaCas9 exhibits a mutagenic activity that is similar to that of SpCas9 [45]. When the cleavage activity of SpCas9, SaCas9, and a

Class 2 type V nuclease FnCpf1 was compared, SaCas9 achieved a higher level of activity than the other two enzymes for the same target sequence in human cells [150]. The authors acknowledge that differences in cleavage activity between Cas9 and Cpf1 may be due to the different DNA ends generated (blunt ends vs cohesive ends), yet their finding still suggest that SaCas9 may be the best choice for gene editing in human cells [150].

The longer PAM sequence of SaCas9 limits its targetability. Kleinstiver et al. manipulated the enzyme to produce a variant that recognizes a NNNRRT PAM (KKH SaCas9, E782K/N968K/R1015H) and shows WT levels of efficiency and specificity [46]. Qin et al. developed a series of mutagenic and base-editing tools derived from SaCas9, including a conventional nickase (nSaCas9, D10A), PAM-requirement altered nickases (nSaKKHCas9, D10A), and base editors (Sa-BE3, SaKKH-BE3), demonstrating that engineering strategies originally developed for SpCas9 can be optimized for orthologous Cas proteins [151]. As noted previously, the large size of the RFN complex and the necessity for dual complexes makes delivery of the conventional SpCas9 system a challenge. RFN-SaCas9 has since been developed to address this size concern, demonstrating efficient modification of the genome when sgRNA pars were spaced 18 bp or 25–28 bp apart [125]. Perhaps more interestingly, it was also found that RFN orthologs can heterodimerize to modify genes, suggesting that future applications will include mixing and matching orthologs [125]. Furthermore, although SaCas9 can fit into a single AAV, a split-nuclease was engineered which could potentially further improve delivery [152]. Chen et al. took advantage of the orthogonal nature of SaCas9 and SpCas9 to produce a live two-color CRISPR imaging system that resolves genomic loci within <300 kn in HEK293 cells [153].

Although double-stranded binding and cleavage is a defining feature of type II Cas proteins, recent work has shown the Cas9s have the ability to recognize and cleave single-stranded RNA (ssRNA) independent of a PAM sequence [154]. SaCas9 demonstrates this natural RNA-guided, RNase activity—it conferred protection against an RNA phage in vivo and repressed gene expression in *E. coli* [154]. This work is a promising foundation for use of Cas9 as a tool to modify both DNA and RNA, particularly as therapeutic targeting RNA viruses.

Streptococcus thermophilus *Cas9*

S. thermophilus was the first bacterium in which the CRISPR-Cas system was discovered [155]. Cas9 has been derived from two of its CRISPR loci, CRISPR1 (St1Cas9) and CRISPR3 (St3Cas9). Both St1Cas9 and St2Cas9 have more stringent PAM requirements than SpCas9 (NNAGAAW and NGGNG, respectively) and recognize a target 19–20 nt in length (Fig. 2.4) [156,157]. St1Cas9 and St3Cas9 have shown cleavage activity in eukaryotes [158], mammalian cells [8], mouse zygotes [159], and human cells [8,19,45,55,158]. Glemzaite et al. showed that the sgRNA–St3Cas9 complex could be delivered directly into HEK293 cells, a potential strategy for editing of human cells in vitro without cloning procedures [49].

Studies comparing St1Cas9 activity to other orthologs have garnered mixed results. compared to *Neisseria meningitides* Cas9 (NmCas9), *Treponema denticola* Cas9, and SpCas9, St1Cas9 was found to produce higher rates of homologous recombination [55]. However, a comparison among St1Cas9, SaCas9, and SpCas9 at two different human targets reveals that St1Cas9 has lower levels of activity [45]. Other studies have demonstrated that St1Cas9 produced comparable frequencies of NHEJ when compared to SpCas9 [48,55,158,160]. An exploration of the orthogonal properties of St1, when compared to NmCas9 and SpCas9, demonstrated that the three enzymes are capable of targeting distinct, nonoverlapping sequences of DNA [55]. Although certain PAMs enabled both St1Cas9 and SpCas9 to target the same sequence, cleavage only occurred when the enzyme was paired with its appropriately sized sgRNA, highlighting the unique specificities of differing Cas9 orthologs [55]. The ability of multiple Cas9 orthologs, even those from the same species such as *S. thermophilus,* to function independently within a given organism is a promising phenomenon for designing coexisting CRISPR-Cas strategies [161].

Neisseria meningitides *Cas9*

The Cas9 of *N. meningitides* has also been explored as an approach to not only improve CRISPR-Cas delivery due to its small size (1109 aa) but also enhance specificity through its complex PAM sequence (NNNNGATT) and longer optimal target sequence (24 nt) (Fig. 2.4) [51,162,163]. NmCas9 has generated homologous recombination in human cells including pluripotent stem cells [50,55,163]. In vivo delivery of NmCas9 via AAV vector to Pcsk9- and Hpd-deficient mice resulted in efficient genomic editing, as measured by decreased level of serum cholesterol and no noted abnormalities or liver damage [164]. Although NmCas9 is less efficient than SpCas9, it demonstrates increased specificity with less off-target effects and more successful delivery by viral vector [163,165].

A distinguishing feature of the CRISPR–NmCas9 system is that recently three naturally occurring bacteriophage proteins were found to block NmCas9 target recognition and activity both in vitro [166] and in human cells [162]. This discovery means that we are better able to temporally, spatially, or tissue-specifically regulate enzyme activity. Like other Cas orthologs, a catalytically inactive variant of NmCas9 (dNmCas9) has been engineered, which, when fused with GFP, transcription effector domains, or methylation and demethylation enzymes, demonstrates fluorescent tagging and transcriptional modulation of genome loci [107,110,167]. Six variant PAMs for NMCas9 have recently been validated in human cells, potentially broadening its application [165].

In addition to conventional cleavage of dsDNA, NmCas9 can also direct in vitro cleavage of single-stranded DNA targets in the presence or absence of a tracrRNA and a PAM [168] and cleavage of ssRNA targets independent of a PAM [169]. The DNase tracrRNA-independent activity requires an active HNH domain, divalent metal ions, and a reprogrammable RNA guide with a minimum of 17 contiguous base pairs complimentary to the target ssDNA sequence [168]. The RNase activity also requires a programmable, RNA guide [169]. As discussed in the SaCas9 section, RNase activity of Cas enzymes has promising implications for future editing and therapeutic strategies.

Francisella novicida *Cas9*

F. novicida Cas9 (FnCas9) is the largest of the Cas9 orthologs (1629 aa), and in vitro studies revealed that FnCas9 requires a NGG PAM sequence and 22 nt target sequence [53]. When delivered as a ribonucleoprotein complex to mouse zygotes, it led to efficient, site-specific target editing with no evidence of indel generation [53]. Similar to engineering strategies for other Cas9s, FnCas9 was modified to recognize an alternative PAM sequence of YG (RHAFnCas9, E1369R/E1449H/R1556A) to broaden its target recognition [53].

Investigation of FnCas9's function has revealed that it can target and cleave both positive- and negative-sense ssRNA transcripts independent of its RuvC and HNH endonuclease and PAM domains [170] (Fig. 2.8). A study in human cells infected with hepatitis C virus demonstrated the FnCas9 can inhibit RNA virus by blocking its translation and replication machinery [171]. This study also elucidated that the arginine-rich motif of FnCas9 is necessary for the enzyme's ability to interact with nucleic acid and thus its ability to target ssRNA [171]. Price et al. hypothesize that

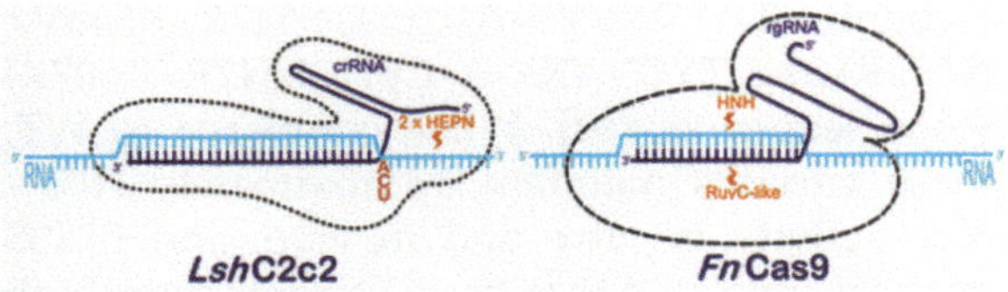

FIGURE 2.8 RNA-guided RNase enzyme systems C2c2 from *Leptotrichia shahii* (*left*) and Cas9 from *Francisella novicida* (*right*). Target ssRNAs (*light blue*) are recognized by complementary regions of crRNA or rgRNA (*dark blue*) in LshC2c2 and FnCas9, respectively. In LshC2c2, cleavage is performed downstream of the protospacer-flanking site (A, C, or U, *red*) by two HEPN domains (*orange*), while in FnCas9 cleavage occurs within the complementary region via a combined activity of HNH and RuvC-like nuclease domains (*orange*). *Source: Reproduced with permission from J. Murovec, Z. Pirc, B. Yang, New variants of CRISPR RNA-guided genome editing enzymes, Plant Biotechnol. J. 15 (8) (2017) 918 [10]. This figure is reproduced under the terms of the Creative Commons Attribution 4.0 International License (http://creativecommons.org/licenses/by/4.0/).*

cytosolic targeting of RNA will limit off-target effects on host DNA [171].

Campylobacter jejuni *Cas9*

One of the smallest Cas9 orthologs (984 aa), *C. jejuni* Cas9, can be easily packaged with its sgRNA in a single AAV vector, making it an attractive candidate for in vivo genome editing [40]. Similar to the other Cas9 orthologs discussed, it has a more stringent PAM requirement of NNNACAC or NNNNRYAC compared to that of SpCas9, and it recognizes a long 22-nt sequence [40]. Study of its crystal structure in complex with sgRNA and target DNA has revealed unique features of the enzyme: unlike other Cas9 orthologs, CjCas9 binds both the target and nontarget DNA strands, and its tracrRNA has a unique triple-helix structure [54]. CjCas9 has successfully modified mammalian cells in vitro and in vivo via AAV delivery [40,172,173]. It has comparable efficiency to SpCAs9 and SaCas9, but was found to be more specific at a number of target sites in mouse muscle and retinal cells [40]. AAV delivery of CjCas9 to retinal cells and myocytes has shown to be a potential therapeutic strategy for treating age-related macular degeneration and Duchenne's muscular dystrophy [40,172,173]. CjCas9 successfully shifts the reading frame in disease-causing loci to restore physiologic function while showing no signs of off-target effects [40,172,173]. Consistent with recently discovered properties of other Cas orthologs, cRNA-guided CjCas9 demonstrates RNase activity, as it can bind and cleave complementary mRNA independent of its HNH and PAM domains [154,174]. In addition to being programmable in vivo, RNase activity predictably improved as matching between the cRNA and RNA target improved [174].

CRISPR-Cas12a (Cpf1) system

The CRISPR-Cas12a (formerly known as Cpf1) system was first characterized in *F. novicida* (FnCpf1) (this locus is distinct from the FnCas9 locus) and is characterized by its signature Class 2 type V effector protein (Fig. 2.4) [58]. This effector lacks a HNH catalytic domain and instead utilizes a RuvC-like nuclease domain to induce staggered DSBs (5 nt overhang) outside of the PAM site unlike Cas9, which has both a HNH and RuvC domain and creates blunt end cuts within the PAM region [9,175,176]. The "sticky" ends generated by Cas12a are beneficial for gene editing, fragment addition, and HDR-replacement applications. Additionally, the enzyme is only about 1200-aa long, a desirable feature for AAV delivery [58]. An analysis of the gene-editing activity of 16 Cpf1 orthologs in human cells revealed that FnCpf1, *Moraxella bovoculi* (Mb3Cpf1), *Lachnospiraceae bacterium* (LbCpf1), and *Acidaminococcus* sp. *BV3L6* (AsCpf1) all demonstrate potential to expand the CRISPR-Cas toolbox [60].

In addition to its unique effector protein architecture and function, the Cas12a system has other distinguishing features when compared to Cas9. First, Cas12a can process its own cRNA with RNase activity [177]. Unlike Cas9 which uses a dual-guide RNA complex (crRNA:tracrRNA), the Cas12a-associated mature cRNA is free of tracrRNA (additionally, the cRNA also resembles that of type I or III due to its inclusion of 5′ repeat sequences and 3′ spacer sequences) [17,178]. T-rich PAM sites are preferred by Cas12 enzymes, whereas Cas9 predominantly targets GC-rich regions of the genome [60].

The Cas12a endoribonuclease is a dual-RNase and -DNase that is ion-dependent and sequence-specific [176,178]. The mature cRNA for Cas12a is 42–44 nt in length, containing a 19 nt repeat and 23–25 nt spacer. At its 5′ end, a single stem loop exists that is favorable for retaining RNase and DNase Cas12a activity [60,177]. Cas12a has the distinct ability to process multiplex crRNA in human cells, particular when crRNA precursor formation is driven by a single RNA Pol II promoter [60,179]. This multiplex activity can expand gene editing and regulation with transfection of a single plasmid. The dual nature of the Cas12a enzyme allows for RNase and DNase activity to be manipulated independently in mammalian cells [61]. Furthermore, mutations in conserved Cas12a RNase residues can disrupt CRISPR array processing but have no influence on RNA binding of the target sequence and/or DNA-cleavage activity [61]. Modifications of the crRNA in the CRISPR-Cas12a system can influence target gene transcription, improve gene editing, and reduce immunogenicity [177,180,181].

In the Cas12a system, the PAM sequence is upstream of the target DNA region on the opposite strand. The canonical PAM for Cas12 orthologs is TTTV, although Mb3Cpf1 demonstrates limited activity with TTN [60] and LbCpf1 shows conformal flexibility with CTTV, TCTC, and TTCV [63,175,182]. Two AsCpf1 variants have been engineered to recognize TATV (RVR-AsCpf1, S542R/K548V/N552R) and TYCV (RR-AsCpf1, S542R/K607R) PAM sequences [62]. When compared to WT, these variants maintain approximately 70%–80% activity in vitro and in vivo, and BLISS analysis indicated that these variants have high specificity at different target sites [62,63]. To further improve the specificity, a K949A mutation was introduced to modify Cpf1's interaction with the noncomplementary strand, decreasing the off-target effects of both variants [62].

Cpf1 is a useful genome-editing approach in bacteria that are unreceptive to the CRISPR-Cas9 system due to the antibacterial toxicity of high-GC content [183,184]. Furthermore, Cas12a applications have shown promise in in vivo studies as a therapeutic strategy. Microinjection of AsCpf1 and LbCpf1 mRNA with corresponding crRNAs in mouse embryos led to 70%–80% on-target mutation frequencies with off-target effects ranging from 16.3% (LbCpf1) to 18.6% (AsCpf1) [185]. Electroporation of recombinant AsCpf1 RNP in mouse embryos displayed 64% hit of target mutations with undetectable levels of off-target effects [185]. AsCpf1 and LbCpf1 efficiently correct Duchene muscular dystrophy-associated loci in both human iPSCs and live mouse models [186]. CRISPR-Cas12a platforms have also shown promise in editing oncogenic mutations like BRAF-V600E. Unlike CRISPR–Cas9 systems which cannot discriminate mutant and normal forms of the BRAF locus, CRISPR–Cas12a systems were able to selectively inactivate the mutant BRAF allele, a finding which is promising for future development of selective inactivation of gain-of-function, oncogenic events [187].

WT Cas12a can induce mutations just as efficiently as wtSpCas9 and SaCas9 at endogenous loci within mammalian cells [45,188]. Examination of the four Cas12a orthologs (LbCpf1, AsCpf1, FnCpf1, and MbCpf1) revealed that LbCpf1 and AsCpf1 are the most efficient of the four in eukaryotic cells when in combination with orthogonal and nonorthogonal crRNAs [58,188,189]. However, FnCpf1 could also be used in human cells as well [59]. Cas12a are extremely sensitive to target mismatching at their 5′ PAM-proximal region. Expectedly, multiple studies have demonstrated that off-target mutation frequency of AsCpf1 and LbCpf1 is substantially lower when compared to that of wtSpCas9 at the same chromosomal sites [45,188,189]. Moreover, GUIDE-seq and deep sequencing analyses have confirmed that Cas12a endonuclease are more specific in human cells than high-fidelity SpCas9 variants [29,31]. In a comparison with Cas9 orthologs (NmCas9, StCas9, and SaCas9), AsCpf1 and LbCpf1 demonstrated greater efficiency at inducing HDR in mouse neuroblastoma cells [190]. Unlike SpCas9 which exhibits a preference for ssODNs of the target sequence to induce HDR, Wang et al. found that two Cpf1 nucleases preferred ssODNs homologous to the nontarget strand [191]. Additionally, Wang et al. developed a systematic approach for selecting Cas9 or Cas12a proteins when designing gene-editing experiments [191].

Cas12a systems have also engineered to not only acquire even better efficiency and specificity but also broaden application use. Chemical modification of both crRNA and Cas12a has had a substantial impact on editing efficiency. Using these two platforms in conjunction results in an almost 300% increase in gene-cutting efficiency in human cells [192]. A DNase-deactivated–AsCpf1 has also shown highly efficient multiplex gene repression in HEK293 cells [61,193]. Recently, Kleinstiver et al. have applied their enhanced, base-editing, and high-fidelity engineering platforms [29] to the Cas12a system, producing enAsCas12a and enAsCas12a-HF1 variants. When delivered as ribonucleoprotein complexes, these variants improve the efficiency of multiplex gene editing, endogenous gene regulation, and C-to-T base editing while reducing off-target activation in human cell lines [194]. A cytidine deaminase-Cas12a base editor (dLbCpf1-BE) has also been developed to catalyze C to T conversion in human cells [134]. Finally, LbCpf1 and AsCpf1 have been engineered to activate transcription through the production Cpf1-p300 and Cpf1-SunTag fusion effectors [195]. Experiments with these fusion effectors produced mixed results with mRNA expression levels only increasing in some of the tested cell lines, suggesting that transcription modification via Cas12a systems can be achieved in a cell-dependent manner [195].

CRISPR-Cas13a (C2c2) system

The CRISPR-Cas13a (formerly known as C2c2) Class 2 type VI system utilizes a single effector protein that, unlike the other orthologs discussed previously, exclusively targets RNA via two HEPN (higher eukaryote and prokaryote nucleotide-binding) RNase domains (Fig. 2.8) [66]. Similar to Cas12a, Cas13a is directed to its target by a single cRNA with an optimal length of 64–66 nt and has a preference for sequence with C, A, and U flanking the 3′ end of a 24–44 nt target sequence [66,196]. The crRNA requires target homology and the secondary structure of its repeat stem in vitro. This system was confirmed by expression of *Leptotrichia shahii* (LsCas13a) and a corresponding crRNA in *E. coli*, which subsequently targeted bacteriophage ssRNA and protected the bacterial cell from infection [66]. Additionally, these enzymes can gain nonspecific RNase activity, leading to collateral RNA cleavage [66]. It has been hypothesized that this nonspecific activity, as demonstrated by *Leptotrichia buccalis* Cas13a (LbCas13a) in HeLa cells, is activated by the presence of a specific target RNA [198].

This enzyme shows promise as an experimental tool to ablate target RNA transcripts in the absence of genomic cleavage or, if catalytically deactivated, as a tool for localizing and imaging RNA populations in a cell. In a test to knockdown KRAB, CXC4, and PPIB endogenous genes at the posttranslational level in HEK293 cells, *Letpotrichia wadei* (LwaCas13a) disrupted transcript activity to a degree that is comparable to RNA interference experiments [196]. A reprogrammable, RNA-binding protein for modification of mammalian cell transcripts was produced by deactivating the LwcCas13a HFEN residues (dLwcCas13a) [196]. Of note, recently a novel RNA-targeting CRISPR effector, Cas13d, has been described to knockdown endogenous mRNAs and long noncoding RNAs involved with cancer, cell signaling, and epigenetic modulation [189] while being about 20% smaller than Cas13a effectors, making it an attractive tool for in vivo editing. Recently, the base-editing approach has also been applied to the Cas13 system. A deactivated-variant of another Class 2 type VI protein, Cas13b, was fused to a deaminase domain (ADAR2) to create a system that performs A to I base conversation and enables single nucleotide editing without RNA cutting [197]. Additionally NmCas9, SpCas9, and Cas13a have orthogonal RNAs, a key feature that could be used in the future to achieve distinct but multiplexed RNA-targeting applications [55,66]. A new RNA-targeting strategy called SHERLOCK (specific high detection of DNA and RNA) uses Cas13a to detect low-frequency mutations in ssDNA and ssRNA to levels of 1 and 2 aM, respectively, showing promise for identification of somatic cancer mutations at conventionally indiscernible levels [198].

Conclusion

In the few years it has been since the seminal *S. pyogenes* CRISPR/Cas9 was discovered to cut and edit mammalian genomes, a deluge of scientific exploration has revealed the wealth of Cas enzymes at our disposal. Analysis of bacterial sequence databases has been a productive technique for discovering novel enzymes, and structural studies have provided a means of adjusting properties of SpCas9 to increase fidelity or broaden its targets across the genome. Challenges of SpCas9-delivery in vivo prompted the exploration of differently sized Cas orthologs and systems. Ranging in size, activity, and specificity, these described effectors enable us with a diverse set of parameters to modify, regulate, label, disrupt, activate, and explore nucleic acid complexes. The diversity of this toolbox has great potential to facilitate future gene therapy and treatment applications.

References

[1] K.S. Makarova, Y.I. Wolf, O.S. Alkhnbashi, F. Costa, S. A. Shah, S.J. Saunders, et al., An updated evolutionary classification of CRISPR–Cas systems, Nat. Rev. Microbiol. 13 (2015) 722–736. Available from: https://doi.org/10.1038/nrmicro3569.

[2] K.S. Makarova, E.V. Koonin, Annotation and Classification of CRISPR-Cas Systems, CRISPR, Springer, New York, NY, 2015, pp. 47–75. Available from: https://doi.org/10.1007/978-1-4939-2687-9_4.

[3] K.S. Makarova, Y.I. Wolf, E.V. Koonin, The basic building blocks and evolution of CRISPR–Cas systems, Biochem. Soc. Trans. 41 (2013) 1392–1400. Available from: https://doi.org/10.1042/BST20130038.

[4] M. Jinek, K. Chylinski, I. Fonfara, M. Hauer, J.A. Doudna, E. Charpentier, A programmable dual-RNA-guided DNA endonuclease in adaptive bacterial immunity, Science 337 (2012) 816–821. Available from: https://doi.org/10.1126/science.1225829.

[5] G. Gasiunas, R. Barrangou, P. Horvath, V. Siksnys, Cas9-crRNA ribonucleoprotein complex mediates specific DNA cleavage for adaptive immunity in bacteria, Proc. Natl Acad. Sci. U.S.A. 109 (2012) E2579–E2586. Available from: https://doi.org/10.1073/pnas.1208507109.

[6] M. Jinek, A. East, A. Cheng, S. Lin, E. Ma, J. Doudna, RNA-programmed genome editing in human cells, Elife 2 (2013) e00471. Available from: https://doi.org/10.7554/eLife.00471.

[7] S.W. Cho, S. Kim, J.M. Kim, J.-S. Kim, Targeted genome engineering in human cells with the Cas9 RNA-guided endonuclease, Nat. Biotechnol. 31 (2013) 230–232. Available from: https://doi.org/10.1038/nbt.2507.

[8] L. Cong, F.A. Ran, D. Cox, S. Lin, R. Barretto, N. Habib, et al., Multiplex genome engineering using CRISPR/Cas systems, Science 339 (2013) 819–823. Available from: https://doi.org/10.1126/science.1231143.

[9] P. Mali, L. Yang, K.M. Esvelt, J. Aach, M. Guell, J.E. DiCarlo, et al., RNA-guided human genome engineering via Cas9, Science 339 (2013) 823–826. Available from: https://doi.org/10.1126/science.1232033.

[10] J. Murovec, Ž. Pirc, B. Yang, New variants of CRISPR RNA-guided genome editing enzymes, Plant. Biotechnol. J. 15 (2017) 917–926. Available from: https://doi.org/10.1111/pbi.12736.

[11] M.R. Lieber, The mechanism of double-strand DNA break repair by the nonhomologous DNA end joining pathway, Annu. Rev. Biochem. 79 (2010) 181–211. Available from: https://doi.org/10.1146/annurev.biochem.052308.093131.

[12] M. McVey, S.E. Lee, MMEJ repair of double-strand breaks (director's cut): deleted sequences and alternative endings, Trends Genet. 24 (2008) 529–538. Available from: https://doi.org/10.1016/j.tig.2008.08.007.

[13] S. Sinha, F. Li, D. Villarreal, J.H. Shim, S. Yoon, K. Myung, et al., Microhomology-mediated end joining induces hypermutagenesis at breakpoint junctions, PLoS Genet. 13 (2017) e1006714. Available from: https://doi.org/10.1371/journal.pgen.1006714.

[14] E.C. Greene, DNA sequence alignment during homologous recombination, J. Biol. Chem. 291 (2016) 11572–11580. Available from: https://doi.org/10.1074/jbc.R116.724807.

[15] V.T. Chu, T. Weber, R. Graf, T. Sommermann, K. Petsch, U. Sack, et al., Efficient generation of Rosa26 knock-in mice using CRISPR/Cas9 in C57BL/6 zygotes, BMC Biotechnol. 16 (2016). Available from: https://doi.org/10.1186/s12896-016-0234-4.

[16] Y. Ding, H. Li, L.-L. Chen, K. Xie, Recent advances in genome editing using CRISPR/Cas9, Front. Plant. Sci. 7 (2016). Available from: https://doi.org/10.3389/fpls.2016.00703.

[17] E. Deltcheva, K. Chylinski, C.M. Sharma, K. Gonzales, Y. Chao, Z.A. Pirzada, et al., CRISPR RNA maturation by trans-encoded small RNA and host factor RNase III, Nature 471 (2011) 602–607. Available from: https://doi.org/10.1038/nature09886.

[18] J.D. Sander, J.K. Joung, CRISPR-Cas systems for editing, regulating and targeting genomes, Nat. Biotechnol. 32 (2014) 347–355. Available from: https://doi.org/10.1038/nbt.2842.

[19] B.P. Kleinstiver, M.S. Prew, S.Q. Tsai, V.V. Topkar, N.T. Nguyen, Z. Zheng, et al., Engineered CRISPR-Cas9 nucleases with altered PAM specificities, Nature 523 (2015) 481–485. Available from: https://doi.org/10.1038/nature14592.

[20] D.-J.J. Truong, K. Kühner, R. Kühn, S. Werfel, S. Engelhardt, W. Wurst, et al., Development of an intein-mediated split–Cas9 system for gene therapy, Nucleic Acids Res. 43 (2015) 6450–6458. Available from: https://doi.org/10.1093/nar/gkv601.

[21] C.L. Xu, M.Z.C. Ruan, V.B. Mahajan, S.H. Tsang, Viral delivery systems for CRISPR, Viruses 11 (2019). Available from: https://doi.org/10.3390/v11010028.

[22] P. Mali, J. Aach, P.B. Stranges, K.M. Esvelt, M. Moosburner, S. Kosuri, et al., CAS9 transcriptional activators for target specificity screening and paired nickases for cooperative genome engineering, Nat. Biotechnol. 31 (2013) 833–838. Available from: https://doi.org/10.1038/nbt.2675.

[23] F.A. Ran, P.D. Hsu, C.-Y. Lin, J.S. Gootenberg, S. Konermann, A. Trevino, et al., Double nicking by RNA-guided CRISPR Cas9 for enhanced genome editing specificity, Cell 154 (2013) 1380–1389. Available from: https://doi.org/10.1016/j.cell.2013.08.021.

[24] F.A. Ran, P.D. Hsu, J. Wright, V. Agarwala, D.A. Scott, F. Zhang, Genome engineering using the CRISPR-Cas9 system, Nat. Protoc. 8 (2013) 2281–2308. Available from: https://doi.org/10.1038/nprot.2013.143.

[25] S.W. Cho, S. Kim, Y. Kim, J. Kweon, H.S. Kim, S. Bae, et al., Analysis of off-target effects of CRISPR/Cas-derived RNA-guided endonucleases and nickases, Genome Res. 24 (2014) 132–141. Available from: https://doi.org/10.1101/gr.162339.113.

[26] C. Anders, K. Bargsten, M. Jinek, Structural plasticity of PAM recognition by engineered variants of the RNA-guided endonuclease Cas9, Mol. Cell 61 (2016) 895–902. Available from: https://doi.org/10.1016/j.molcel.2016.02.020.

[27] J.H. Hu, S.M. Miller, M.H. Geurts, W. Tang, L. Chen, N. Sun, et al., Evolved Cas9 variants with broad PAM compatibility and high DNA specificity, Nature 556 (2018) 57–63. Available from: https://doi.org/10.1038/nature26155.

[28] H. Nishimasu, L. Cong, W.X. Yan, F.A. Ran, B. Zetsche, Y. Li, A. Kurabayashi, R. Ishitani, F. Zhang, O. Nureki, Crystal Structure of Staphylococcus aureus Cas9, Cell 162 (5) (2015) 1113–1126. Available from: https://doi.org/10.1016/j.cell.2015.08.007. PMID: 26317473; PMCID: PMC4670267.

[29] B.P. Kleinstiver, V. Pattanayak, M.S. Prew, S.Q. Tsai, N. Nguyen, Z. Zheng, et al., High-fidelity CRISPR-Cas9 variants with undetectable genome-wide off-targets, Nature 529 (2016) 490–495. Available from: https://doi.org/10.1038/nature16526.

[30] B.P. Kleinstiver, S.Q. Tsai, M.S. Prew, N.T. Nguyen, M.M. Welch, J.M. Lopez, et al., Genome-wide specificities of CRISPR-Cas Cpf1 nucleases in human cells,

Nat. Biotechnol. 34 (2016) 869–874. Available from: https://doi.org/10.1038/nbt.3620.
[31] I.M. Slaymaker, L. Gao, B. Zetsche, D.A. Scott, W.X. Yan, F. Zhang, Rationally engineered Cas9 nucleases with improved specificity, Science 351 (2016) 84–88. Available from: https://doi.org/10.1126/science.aad5227.
[32] D. Bikard, W. Jiang, P. Samai, A. Hochschild, F. Zhang, L.A. Marraffini, Programmable repression and activation of bacterial gene expression using an engineered CRISPR-Cas system, Nucleic Acids Res. 41 (2013) 7429–7437. Available from: https://doi.org/10.1093/nar/gkt520.
[33] L.S. Qi, M.H. Larson, L.A. Gilbert, J.A. Doudna, J.S. Weissman, A.P. Arkin, et al., Repurposing CRISPR as an RNA-guided platform for sequence-specific control of gene expression, Cell 152 (2013) 1173–1183. Available from: https://doi.org/10.1016/j.cell.2013.02.022.
[34] S.Q. Tsai, N. Wyvekens, C. Khayter, J.A. Foden, V. Thapar, D. Reyon, et al., Dimeric CRISPR RNA-guided FokI nucleases for highly specific genome editing, Nat. Biotechnol. 32 (2014) 569–576. Available from: https://doi.org/10.1038/nbt.2908.
[35] J.P. Guilinger, D.B. Thompson, D.R. Liu, Fusion of catalytically inactive Cas9 to FokI nuclease improves the specificity of genome modification, Nat. Biotechnol. 32 (2014) 577–582. Available from: https://doi.org/10.1038/nbt.2909.
[36] X.S. Liu, H. Wu, X. Ji, Y. Stelzer, X. Wu, S. Czauderna, et al., Editing DNA methylation in the mammalian genome, Cell 167 (1) (2016) 233–247. Available from: https://doi.org/10.1016/j.cell.2016.08.056. PMID: 27662091; PMCID: PMC5062609.
[37] A.C. Komor, Y.B. Kim, M.S. Packer, J.A. Zuris, D.R. Liu, Programmable editing of a target base in genomic DNA without double-stranded DNA cleavage, Nature 533 (2016) 420–424. Available from: https://doi.org/10.1038/nature17946.
[38] K. Nishida, T. Arazoe, N. Yachie, S. Banno, M. Kakimoto, M. Tabata, et al., Targeted nucleotide editing using hybrid prokaryotic and vertebrate adaptive immune systems, Science 353 (2016). Available from: https://doi.org/10.1126/science.aaf8729.
[39] K. Kim, S.-M. Ryu, S.-T. Kim, G. Baek, D. Kim, K. Lim, et al., Highly efficient RNA-guided base editing in mouse embryos, Nat. Biotechnol. 35 (2017) 435–437. Available from: https://doi.org/10.1038/nbt.3816.
[40] E. Kim, T. Koo, S.W. Park, D. Kim, K. Kim, H.-Y. Cho, et al., In vivo genome editing with a small Cas9 orthologue derived from *Campylobacter jejuni*, Nat. Commun. 8 (2017) 14500. Available from: https://doi.org/10.1038/ncomms14500.
[41] N.M. Gaudelli, A.C. Komor, H.A. Rees, M.S. Packer, A.H. Badran, D.I. Bryson, et al., Programmable base editing of A•T to G•C in genomic DNA without DNA cleavage, Nature 551 (2017) 464–471. Available from: https://doi.org/10.1038/nature24644.
[42] A. Cebrian-Serrano, B. Davies, CRISPR-Cas orthologues and variants: optimizing the repertoire, specificity and delivery of genome engineering tools, Mamm. Genome 28 (2017) 247–261. Available from: https://doi.org/10.1007/s00335-017-9697-4.
[43] D.J.W. Brocken, M. Tark-Dame, R.T. Dame, dCas9: a versatile tool for epigenome editing, Curr. Issues Mol. Biol. 26 (2018) 15–32. Available from: https://doi.org/10.21775/cimb.026.015.
[44] K. Liu, C. Petree, T. Requena, P. Varshney, G.K. Varshney, Expanding the CRISPR toolbox in Zebrafish for studying development and disease, Front. Cell Dev. Biol. 7 (2019). Available from: https://doi.org/10.3389/fcell.2019.00013.
[45] F.A. Ran, L. Cong, W.X. Yan, D.A. Scott, J.S. Gootenberg, A.J. Kriz, et al., In vivo genome editing using *Staphylococcus aureus* Cas9, Nature 520 (2015) 186–191. Available from: https://doi.org/10.1038/nature14299.
[46] B.P. Kleinstiver, M.S. Prew, S.Q. Tsai, N.T. Nguyen, V. V. Topkar, Z. Zheng, et al., Broadening the targeting range of *Staphylococcus aureus* CRISPR-Cas9 by modifying PAM recognition, Nat. Biotechnol. 33 (2015) 1293–1298. Available from: https://doi.org/10.1038/nbt.3404.
[47] Friedland, A. E., Baral, R., Singhal, P., Loveluck, K., Shen, S., Sanchez, M., et al. (2015). Characterization of Staphylococcus aureus Cas9: a smaller Cas9 for all-in-one adeno-associated virus delivery and paired nickase applications. Genome Biol. 16 (2015) 257. Available from: https://doi.org/10.1186/s13059-015-0817-8.
[48] M. Müller, C.M. Lee, G. Gasiunas, T.H. Davis, T.J. Cradick, V. Siksnys, et al., *Streptococcus thermophilus* CRISPR-Cas9 systems enable specific editing of the human genome, Mol. Ther. 24 (2016) 636–644. Available from: https://doi.org/10.1038/mt.2015.218.
[49] M. Glemzaite, E. Balciunaite, T. Karvelis, G. Gasiunas, M.M. Grusyte, G. Alzbutas, et al., Targeted gene editing by transfection of in vitro reconstituted *Streptococcus thermophilus* Cas9 nuclease complex, RNA Biol. 12 (2015) 1–4. Available from: https://doi.org/10.1080/15476286.2015.1017209.
[50] Z. Hou, Y. Zhang, N.E. Propson, S.E. Howden, L.-F. Chu, E.J. Sontheimer, et al., Efficient genome engineering in human pluripotent stem cells using Cas9 from *Neisseria meningitidis*, Proc. Natl Acad. Sci. U.S.A. 110 (2013) 15644–15649. Available from: https://doi.org/10.1073/pnas.1313587110.
[51] I. Fonfara, A. Le Rhun, K. Chylinski, K.S. Makarova, A.-L. Lécrivain, J. Bzdrenga, et al., Phylogeny of Cas9 determines functional exchangeability of dual-RNA

and Cas9 among orthologous type II CRISPR-Cas systems, Nucleic Acids Res. 42 (2014) 2577–2590. Available from: https://doi.org/10.1093/nar/gkt1074.

[52] S. Hirano, H. Nishimasu, R. Ishitani, O. Nureki, Structural basis for the altered PAM specificities of engineered CRISPR-Cas9, Mol. Cell 61 (2016) 886–894. Available from: https://doi.org/10.1016/j.molcel.2016.02.018.

[53] H. Hirano, J.S. Gootenberg, T. Horii, O.O. Abudayyeh, M. Kimura, P.D. Hsu, et al., Structure and engineering of *Francisella novicida* Cas9, Cell 164 (2016) 950–961. Available from: https://doi.org/10.1016/j.cell.2016.01.039.

[54] M. Yamada, Y. Watanabe, J.S. Gootenberg, H. Hirano, F.A. Ran, T. Nakane, et al., Crystal structure of the minimal Cas9 from *Campylobacter jejuni* reveals the molecular diversity in the CRISPR-Cas9 systems, Mol. Cell 65 (2017) 1109–1121. Available from: https://doi.org/10.1016/j.molcel.2017.02.007. e3.

[55] K.M. Esvelt, P. Mali, J.L. Braff, M. Moosburner, S.J. Yaung, G.M. Church, Orthogonal Cas9 proteins for RNA-guided gene regulation and editing, Nat. Methods 10 (2013) 1116–1121. Available from: https://doi.org/10.1038/nmeth.2681.

[56] P. Chatterjee, N. Jakimo, J.M. Jacobson, Minimal PAM specificity of a highly similar SpCas9 ortholog, Sci. Adv. 4 (10) (2018) eaau0766. Available from: https://doi.org/10.1126/sciadv.aau0766. PMID: 30397647; PMCID: PMC6200363.

[57] T. Gong, B. Tang, X. Zhou, J. Zeng, M. Lu, X. Guo, et al., Genome editing in Streptococcus mutans through self-targeting CRISPR arrays, Mol. Oral Microbiol. 33 (6) (2018) 440–449. Available from: https://doi.org/10.1111/omi.12247. Epub 2018 Oct 24. PMID: 30329221.

[58] B. Zetsche, J.S. Gootenberg, O.O. Abudayyeh, I.M. Slaymaker, K.S. Makarova, P. Essletzbichler, et al., Cpf1 is a single RNA-guided endonuclease of a Class 2 CRISPR-Cas system, Cell 163 (2015) 759–771. Available from: https://doi.org/10.1016/j.cell.2015.09.038.

[59] M. Tu, L. Lin, Y. Cheng, X. He, H. Sun, H. Xie, et al., A 'new lease of life': FnCpf1 possesses DNA cleavage activity for genome editing in human cells, Nucleic Acids Res. 45 (2017) 11295–11304. Available from: https://doi.org/10.1093/nar/gkx783.

[60] B. Zetsche, J. Strecker, O.O. Abudayyeh, J.S. Gootenberg, D.A. Scott, F. Zhang, A survey of genome editing activity for 16 Cpf1 orthologs, BioRxiv (2017). Available from: https://doi.org/10.1101/134015.

[61] B. Zetsche, M. Heidenreich, P. Mohanraju, I. Fedorova, J. Kneppers, E.M. DeGennaro, et al., Multiplex gene editing by CRISPR-Cpf1 through autonomous processing of a single crRNA array, Nat. Biotechnol. 35 (2017) 31–34. Available from: https://doi.org/10.1038/nbt.3737.

[62] L. Gao, D.B.T. Cox, W.X. Yan, J.C. Manteiga, M.W. Schneider, T. Yamano, et al., Engineered Cpf1 variants with altered PAM specificities, Nat. Biotechnol. 35 (2017) 789–792. Available from: https://doi.org/10.1038/nbt.3900.

[63] H. Nishimasu, T. Yamano, L. Gao, F. Zhang, R. Ishitani, O. Nureki, Structural Basis for the altered PAM recognition by engineered CRISPR-Cpf1, Mol. Cell 67 (2017) 139–147. Available from: https://doi.org/10.1016/j.molcel.2017.04.019. e2.

[64] E. Tóth, É. Varga, P. István Kulcsár, V. Kocsis-Jutka, S. Laura Krausz, A. Nyeste, et al., Improved LbCas12a variants with altered PAM specificities further broaden the genome targeting range of Cas12a nucleases, Nucleic Acids Res 48 (7) (2020) 3722–3733. Available from: https://doi.org/10.1093/nar/gkaa110.

[65] A. East-Seletsky, M.R. O'Connell, S.C. Knight, D. Burstein, J.H. Cate, R. Tjian, et al., Two distinct RNase activities of CRISPR-C2c2 enable guide-RNA processing and RNA detection, Nature 538 (7624) (2016) 270–273. Available from: https://doi.org/10.1038/nature19802. Epub 2016 Sep 26. PMID: 27669025; PMCID: PMC5576363.

[66] O.O. Abudayyeh, J.S. Gootenberg, S. Konermann, J. Joung, I.M. Slaymaker, D.B.T. Cox, et al., C2c2 is a single-component programmable RNA-guided RNA-targeting CRISPR effector, Science 353 (2016). Available from: https://doi.org/10.1126/science.aaf5573.

[67] R. Eggenschwiler, M. Moslem, M.S. Fráguas, M. Galla, O. Papp, M. Naujock, et al., Improved bi-allelic modification of a transcriptionally silent locus in patient-derived iPSC by Cas9 nickase, Sci. Rep. 6 (2016). Available from: https://doi.org/10.1038/srep38198.

[68] G.L. Dianov, U. Hübscher, Mammalian base excision repair: the forgotten Archangel, Nucleic Acids Res. 41 (2013) 3483–3490. Available from: https://doi.org/10.1093/nar/gkt076.

[69] X.A. Ge, C.P. Hunter, Efficient homologous recombination in mice using long single stranded DNA and CRISPR Cas9 nickase, G3 (Bethesda) 9 (2018) 281–286. Available from: https://doi.org/10.1534/g3.118.200758.

[70] B. Shen, W. Zhang, J. Zhang, J. Zhou, J. Wang, L. Chen, et al., Efficient genome modification by CRISPR-Cas9 nickase with minimal off-target effects, Nat. Methods 11 (2014) 399–402. Available from: https://doi.org/10.1038/nmeth.2857.

[71] C.J. Jung, J. Zhang, E. Trenchard, K.C. Lloyd, D.B. West, B. Rosen, et al., Efficient gene targeting in mouse zygotes mediated by CRISPR/Cas9-protein, Transgen. Res. 26 (2017) 263–277. Available from: https://doi.org/10.1007/s11248-016-9998-5.

[72] J. Wu, S.D. Hunt, H. Xue, Y. Liu, R. Darabi, Generation and validation of PAX7 reporter lines from human iPS cells using CRISPR/Cas9 technology, Stem Cell Res. 16 (2016) 220–228. Available from: https://doi.org/10.1016/j.scr.2016.01.003.

[73] M.J. Osborn, R. Gabriel, B.R. Webber, A.P. DeFeo, A. N. McElroy, J. Jarjour, et al., Fanconi anemia gene editing by the CRISPR/Cas9 system, Hum. Gene Ther. 26 (2015) 114–126. Available from: https://doi.org/10.1089/hum.2014.111.

[74] H.J. van de Vrugt, T. Harmsen, J. Riepsaame, G. Alexantya, S.E. van Mil, Y. de Vries, et al., Effective CRISPR/Cas9-mediated correction of a Fanconi anemia defect by error-prone end joining or templated repair, Sci. Rep. 9 (2019). Available from: https://doi.org/10.1038/s41598-018-36506-w.

[75] Y. Wang, J. Zhao, N. Duan, W. Liu, Y. Zhang, M. Zhou, et al., Paired CRISPR/Cas9 nickases mediate efficient site-specific integration of F9 into rDNA locus of mouse ESCs, Int. J. Mol. Sci. 19 (2018). Available from: https://doi.org/10.3390/ijms19103035.

[76] Y. Wang, L. Hao, H. Wang, K. Santostefano, A. Thapa, J. Cleary, et al., Therapeutic genome editing for myotonic dystrophy type 1 using CRISPR/Cas9, Mol. Ther. 26 (2018) 2617–2630. Available from: https://doi.org/10.1016/j.ymthe.2018.09.003.

[77] T. Sakuma, K. Masaki, H. Abe-Chayama, K. Mochida, T. Yamamoto, K. Chayama, Highly multiplexed CRISPR-Cas9-nuclease and Cas9-nickase vectors for inactivation of hepatitis B virus, Genes. Cell 21 (2016) 1253–1262. Available from: https://doi.org/10.1111/gtc.12437.

[78] K.D. Rysenkova, E.V. Semina, M.N. Karagyaur, A.A. Shmakova, D.T. Dyikanov, P.A. Vasiluev, et al., CRISPR/Cas9 nickase mediated targeting of urokinase receptor gene inhibits neuroblastoma cell proliferation, Oncotarget 9 (2018) 29414–29430. Available from: https://doi.org/10.18632/oncotarget.25647.

[79] L. Ji, G. Zhao, P. Zhang, W. Huo, P. Dong, H. Watari, et al., Knockout of MTF1 inhibits the epithelial to mesenchymal transition in ovarian cancer cells, J. Cancer 9 (2018) 4578–4585. Available from: https://doi.org/10.7150/jca.28040.

[80] P. Bialk, N. Rivera-Torres, B. Strouse, E.B. Kmiec, Regulation of gene editing activity directed by single-stranded oligonucleotides and CRISPR/Cas9 systems, PLoS One 10 (2015) e0129308. Available from: https://doi.org/10.1371/journal.pone.0129308.

[81] Y. Miyaoka, J.R. Berman, S.B. Cooper, S.J. Mayerl, A.H. Chan, B. Zhang, et al., Systematic quantification of HDR and NHEJ reveals effects of locus, nuclease, and cell type on genome-editing, Sci. Rep. 6 (2016). Available from: https://doi.org/10.1038/srep23549.

[82] H. Nishimasu, F.A. Ran, P.D. Hsu, S. Konermann, S. Shehata, N. Dohmae, et al., Crystal structure of Cas9 in complex with guide RNA and target DNA, Cell 156 (2014) 935–949. Available from: https://doi.org/10.1016/j.cell.2014.02.001.

[83] C. Anders, O. Niewoehner, A. Duerst, M. Jinek, Structural basis of PAM-dependent target DNA recognition by the Cas9 endonuclease, Nature 513 (2014) 569–573. Available from: https://doi.org/10.1038/nature13579.

[84] A. Idoko-Akoh, L. Taylor, H.M. Sang, M.J. McGrew, High fidelity CRISPR/Cas9 increases precise monoallelic and biallelic editing events in primordial germ cells, Sci. Rep. 8 (2018). Available from: https://doi.org/10.1038/s41598-018-33244-x.

[85] F. Chen, X. Ding, Y. Feng, T. Seebeck, Y. Jiang, G.D. Davis, Targeted activation of diverse CRISPR-Cas systems for mammalian genome editing via proximal CRISPR targeting, Nat. Commun. 8 (2017) 14958. Available from: https://doi.org/10.1038/ncomms14958.

[86] A. Casini, M. Olivieri, G. Petris, C. Montagna, G. Reginato, G. Maule, et al., A highly specific SpCas9 variant is identified by in vivo screening in yeast, Nat. Biotechnol. 36 (2018) 265–271. Available from: https://doi.org/10.1038/nbt.4066.

[87] C.A. Vakulskas, D.P. Dever, G.R. Rettig, R. Turk, A.M. Jacobi, M.A. Collingwood, et al., A high-fidelity Cas9 mutant delivered as a ribonucleoprotein complex enables efficient gene editing in human haematopoietic stem and progenitor cells, Nat. Med. 24 (2018) 1216–1224. Available from: https://doi.org/10.1038/s41591-018-0137-0.

[88] D. Zhang, H. Zhang, T. Li, K. Chen, J.-L. Qiu, C. Gao, Perfectly matched 20-nucleotide guide RNA sequences enable robust genome editing using high-fidelity SpCas9 nucleases, Genome Biol. 18 (2017). Available from: https://doi.org/10.1186/s13059-017-1325-9.

[89] P.I. Kulcsár, A. Tálas, K. Huszár, Z. Ligeti, E. Tóth, N. Weinhardt, et al., Crossing enhanced and high fidelity SpCas9 nucleases to optimize specificity and cleavage, Genome Biol. 18 (2017). Available from: https://doi.org/10.1186/s13059-017-1318-8.

[90] A.C. Komor, K.T. Zhao, M.S. Packer, N.M. Gaudelli, A. L. Waterbury, L.W. Koblan, et al., Improved base excision repair inhibition and bacteriophage Mu Gam protein yields C:G-to-T:A base editors with higher efficiency and product purity, Sci. Advance. 3 (2017). Available from: https://doi.org/10.1126/sciadv.aao4774.

[91] L. Yang, M. Guell, S. Byrne, J.L. Yang, A. De Los Angeles, P. Mali, et al., Optimization of scarless human stem cell genome editing, Nucleic Acids Res. 41 (2013) 9049–9061. Available from: https://doi.org/10.1093/nar/gkt555.

[92] G.M. Findlay, E.A. Boyle, R.J. Hause, J.C. Klein, J. Shendure, Saturation editing of genomic regions by multiplex homology-directed repair, Nature 513 (2014) 120–123. Available from: https://doi.org/10.1038/nature13695.

[93] Y. Zhang, X. Ge, F. Yang, L. Zhang, J. Zheng, X. Tan, et al., Comparison of non-canonical PAMs for CRISPR/Cas9-mediated DNA cleavage in human cells, Sci. Rep. 4 (2015). Available from: https://doi.org/10.1038/srep05405.

[94] L.G.V. Fernandes, L.P. Guaman, S.A. Vasconcellos, M. B. Heinemann, M. Picardeau, A.L.T.O. Nascimento, Gene silencing based on RNA-guided catalytically inactive Cas9 (dCas9): a new tool for genetic engineering in *Leptospira*, Sci. Rep. 9 (2019). Available from: https://doi.org/10.1038/s41598-018-37949-x.

[95] J.M. Peters, A. Colavin, H. Shi, T.L. Czarny, M.H. Larson, S. Wong, et al., A comprehensive, CRISPR-based functional analysis of essential genes in bacteria, Cell 165 (2016) 1493–1506. Available from: https://doi.org/10.1016/j.cell.2016.05.003.

[96] L.A. Gilbert, M.H. Larson, L. Morsut, Z. Liu, G.A. Brar, S.E. Torres, et al., CRISPR-mediated modular RNA-guided regulation of transcription in eukaryotes, Cell 154 (2013) 442–451. Available from: https://doi.org/10.1016/j.cell.2013.06.044.

[97] M.L. Maeder, S.J. Linder, V.M. Cascio, Y. Fu, Q.H. Ho, J.K. Joung, CRISPR RNA-guided activation of endogenous human genes, Nat. Method 10 (2013) 977–979. Available from: https://doi.org/10.1038/nmeth.2598.

[98] P. Perez-Pinera, D.D. Kocak, C.M. Vockley, A.F. Adler, A.M. Kabadi, L.R. Polstein, et al., RNA-guided gene activation by CRISPR-Cas9-based transcription factors, Nat. Method 10 (2013) 973–976. Available from: https://doi.org/10.1038/nmeth.2600.

[99] I.E.B. Lawhorn, J.P. Ferreira, C.L. Wang, Evaluation of sgRNA target sites for CRISPR-mediated repression of TP53, PLoS One 9 (2014). Available from: https://doi.org/10.1371/journal.pone.0113232.

[100] M. Kampmann, CRISPRi and CRISPRa screens in mammalian cells for precision biology and medicine, ACS Chem. Biol. 13 (2018) 406–416. Available from: https://doi.org/10.1021/acschembio.7b00657.

[101] A. Chavez, J. Scheiman, S. Vora, B.W. Pruitt, M. Tuttle, E. Iyer, et al., Highly-efficient Cas9-mediated transcriptional programming, Nat. Method 12 (2015) 326–328. Available from: https://doi.org/10.1038/nmeth.3312.

[102] C. Moses, F. Nugent, C.B. Waryah, B. Garcia-Bloj, A. R. Harvey, P. Blancafort, Activating PTEN tumor suppressor expression with the CRISPR/dCas9 system, Mol. Ther. Nucleic Acids 14 (2018) 287–300. Available from: https://doi.org/10.1016/j.omtn.2018.12.003.

[103] K.E. Savell, S.V. Bach, M.E. Zipperly, J.S. Revanna, N. A. Goska, J.J. Tuscher, et al., CRISPR/dCas9 activation system for robust and specific gene regulation, ENeuro 6 (2019). Available from: https://doi.org/10.1523/ENEURO.0495-18.2019.

[104] J.I. McDonald, H. Celik, L.E. Rois, G. Fishberger, T. Fowler, R. Rees, et al., Reprogrammable CRISPR/Cas9-based system for inducing site-specific DNA methylation, Biol. Open. (2016). Available from: https://doi.org/10.1242/bio.019067.

[105] A. Vojta, P. Dobrinić, V. Tadić, L. Bočkor, P. Korać, B. Julg, et al., Repurposing the CRISPR-Cas9 system for targeted DNA methylation, Nucleic Acids Res. 44 (2016) 5615–5628. Available from: https://doi.org/10.1093/nar/gkw159.

[106] X. Xu, Y. Tao, X. Gao, L. Zhang, X. Li, W. Zou, et al., A CRISPR-based approach for targeted DNA demethylation, Cell Discov. 2 (2016) 16009. Available from: https://doi.org/10.1038/celldisc.2016.9.

[107] N.A. Kearns, H. Pham, B. Tabak, R.M. Genga, N.J. Silverstein, M. Garber, et al., Functional annotation of native enhancers with a Cas9-histone demethylase fusion, Nat. Methods 12 (2015) 401–403. Available from: https://doi.org/10.1038/nmeth.3325.

[108] E.M. Mendenhall, K.E. Williamson, D. Reyon, J.Y. Zou, O. Ram, J.K. Joung, et al., Locus-specific editing of histone modifications at endogenous enhancers using programmable TALE-LSD1 fusions, Nat. Biotechnol. 31 (2013) 1133–1136. Available from: https://doi.org/10.1038/nbt.2701.

[109] A.W. Cheng, N. Jillette, P. Lee, D. Plaskon, Y. Fujiwara, W. Wang, et al., Casilio: a versatile CRISPR-Cas9-Pumilio hybrid for gene regulation and genomic labeling, Cell Res. 26 (2016) 254–257. Available from: https://doi.org/10.1038/cr.2016.3.

[110] I.B. Hilton, A.M. D'Ippolito, C.M. Vockley, P.I. Thakore, G.E. Crawford, T.E. Reddy, et al., Epigenome editing by a CRISPR/Cas9-based acetyltransferase activates genes from promoters and enhancers, Nat. Biotechnol. 33 (2015) 510–517. Available from: https://doi.org/10.1038/nbt.3199.

[111] M.E. Tanenbaum, L.A. Gilbert, L.S. Qi, J.S. Weissman, R.D. Vale, A protein tagging system for signal amplification in gene expression and fluorescence imaging, Cell 159 (2014) 635–646. Available from: https://doi.org/10.1016/j.cell.2014.09.039.

[112] S. Konermann, M.D. Brigham, A.E. Trevino, J. Joung, O.O. Abudayyeh, C. Barcena, et al., Genome-scale transcriptional activation by an engineered CRISPR-Cas9 complex, Nature 517 (2015) 583–588. Available from: https://doi.org/10.1038/nature14136.

[113] J.G. Zalatan, M.E. Lee, R. Almeida, L.A. Gilbert, E. H. Whitehead, M. La Russa, et al., Engineering complex synthetic transcriptional programs with CRISPR RNA scaffolds, Cell 160 (2015) 339–350. Available from: https://doi.org/10.1016/j.cell.2014.11.052.

[114] F. González, Z. Zhu, Z.-D. Shi, K. Lelli, N. Verma, Q. V. Li, et al., An iCRISPR platform for rapid, multiplexable and inducible genome editing in human pluripotent stem cells, Cell Stem Cell 15 (2014) 215–226. Available from: https://doi.org/10.1016/j.stem.2014.05.018.

[115] T. Wang, J.J. Wei, D.M. Sabatini, E.S. Lander, Genetic screens in human cells using the CRISPR/Cas9 system, Science 343 (2014) 80–84. Available from: https://doi.org/10.1126/science.1246981.

[116] L.E. Dow, J. Fisher, K.P. O'Rourke, A. Muley, E.R. Kastenhuber, G. Livshits, et al., Inducible in vivo genome editing with CRISPR/Cas9, Nat. Biotechnol. 33 (2015) 390–394. Available from: https://doi.org/10.1038/nbt.3155.

[117] A.V. Wright, S.H. Sternberg, D.W. Taylor, B.T. Staahl, J.A. Bardales, J.E. Kornfeld, et al., Rational design of a split-Cas9 enzyme complex, Proc. Natl Acad. Sci. U.S.A. 112 (2015) 2984–2989. Available from: https://doi.org/10.1073/pnas.1501698112.

[118] B. Zetche, S.E. Volz, F. Zhang, A. Split, Cas9 architecture for inducible genome editing and transcription modulation, Nat. Biotechnol. 33 (2015) 139–142. Available from: https://doi.org/10.1038/nbt.3149.

[119] Y. Nihongaki, Y. Furuhata, T. Otabe, S. Hasegawa, K. Yoshimoto, M. Sato, CRISPR–Cas9-based photoactivatable transcription systems to induce neuronal differentiation, Nat. Method 14 (2017) 963–966. Available from: https://doi.org/10.1038/nmeth.4430.

[120] Y. Fu, J.A. Foden, C. Khayter, M.L. Maeder, D. Reyon, J.K. Joung, et al., High-frequency off-target mutagenesis induced by CRISPR-Cas nucleases in human cells, Nat. Biotechnol. 31 (2013) 822–826. Available from: https://doi.org/10.1038/nbt.2623.

[121] C. Kuscu, M. Parlak, T. Tufan, J. Yang, K. Szlachta, X. Wei, et al., CRISPR-STOP: gene silencing through base-editing-induced nonsense mutations, Nat. Methods 14 (2017) 710–712. Available from: https://doi.org/10.1038/nmeth.4327.

[122] S.Q. Tsai, Z. Zheng, N.T. Nguyen, M. Liebers, V.V. Topkar, V. Thapar, et al., GUIDE-Seq enables genome-wide profiling of off-target cleavage by CRISPR-Cas nucleases, Nat. Biotechnol. 33 (2015) 187–197. Available from: https://doi.org/10.1038/nbt.3117.

[123] S. Zhang, C.A. Voigt, Engineered dCas9 with reduced toxicity in bacteria: implications for genetic circuit design, Nucleic Acids Res. 46 (2018) 11115–11125. Available from: https://doi.org/10.1093/nar/gky884.

[124] É.S. Vanamee, S. Santagata, A.K. Aggarwal, FokI requires two specific DNA sites for cleavage, J. Mol. Biol. 309 (2001) 69–78. Available from: https://doi.org/10.1006/jmbi.2001.4635.

[125] S. Havlicek, Y. Shen, Y. Alpagu, M.B. Bruntraeger, N.B.M. Zufir, Z.Y. Phuah, et al., Re-engineered RNA-guided FokI-nucleases for improved genome editing in human cells, Mol. Ther. 25 (2017) 342–355. Available from: https://doi.org/10.1016/j.ymthe.2016.11.007.

[126] Y. Fu, J.D. Sander, D. Reyon, V.M. Cascio, J.K. Joung, Improving CRISPR-Cas nuclease specificity using truncated guide RNAs, Nat. Biotechnol. 32 (2014) 279–284. Available from: https://doi.org/10.1038/nbt.2808.

[127] N. Wyvekens, V.V. Topkar, C. Khayter, J.K. Joung, S. Q. Tsai, Dimeric CRISPR RNA-guided FokI-dCas9 nucleases directed by truncated gRNAs for highly specific genome editing, Hum. Gene Ther. 26 (2015) 425–431. Available from: https://doi.org/10.1089/hum.2015.084.

[128] Y. Ma, J. Zhang, W. Yin, Z. Zhang, Y. Song, X. Chang, Targeted AID-mediated mutagenesis (TAM) enables efficient genomic diversification in mammalian cells, Nat. Method 13 (2016) 1029–1035. Available from: https://doi.org/10.1038/nmeth.4027.

[129] Y. Lu, J.-K. Zhu, Precise editing of a target base in the rice genome using a modified CRISPR/Cas9 system, Mol. Plant. 10 (2017) 523–525. Available from: https://doi.org/10.1016/j.molp.2016.11.013.

[130] P. Liang, H. Sun, Y. Sun, X. Zhang, X. Xie, J. Zhang, et al., Effective gene editing by high-fidelity base editor 2 in mouse zygotes, Protein Cell 8 (2017) 601–611. Available from: https://doi.org/10.1007/s13238-017-0418-2.

[131] H.A. Rees, A.C. Komor, W.-H. Yeh, J. Caetano-Lopes, M. Warman, A.S.B. Edge, et al., Improving the DNA specificity and applicability of base editing through protein engineering and protein delivery, Nat. Commun. 8 (2017). Available from: https://doi.org/10.1038/ncomms15790.

[132] Z. Liu, M. Chen, S. Chen, J. Deng, Y. Song, L. Lai, et al., Highly efficient RNA-guided base editing in rabbit, Nat. Commun. 9 (2018) 2717. Available from: https://doi.org/10.1038/s41467-018-05232-2.

[133] W. Jiang, S. Feng, S. Huang, W. Yu, G. Li, G. Yang, et al., BE-PLUS: a new base editing tool with broadened editing window and enhanced fidelity, Cell Res. 28 (2018) 855–861. Available from: https://doi.org/10.1038/s41422-018-0052-4.

[134] Y. Li, S. Ma, L. Sun, T. Zhang, J. Chang, W. Lu, et al., Programmable single and multiplex base-editing in *Bombyx mori* using RNA-guided cytidine deaminases, G3 (Bethesda) 8 (2018) 1701–1709. Available from: https://doi.org/10.1534/g3.118.200134.

[135] Z. Shi, H. Xin, D. Tian, J. Lian, J. Wang, G. Liu, et al., Modeling human point mutation diseases in *Xenopus tropicalis* with a modified CRISPR/Cas9 system, FASEB J. (2019). Available from: https://doi.org/10.1096/fj.201802661R.

[136] M. Wang, J.A. Zuris, F. Meng, H. Rees, S. Sun, P. Deng, et al., Efficient delivery of genome-editing proteins using bioreducible lipid nanoparticles, Proc. Natl Acad. Sci. U.S.A. 113 (2016) 2868–2873. Available from: https://doi.org/10.1073/pnas.1520244113.

[137] J.A. Zuris, D.B. Thompson, Y. Shu, J.P. Guilinger, J.L. Bessen, J.H. Hu, et al., Efficient delivery of genome-editing proteins in vitro and in vivo, Nat. Biotechnol. 33 (2015) 73–80. Available from: https://doi.org/10.1038/nbt.3081.

[138] X. Liang, J. Potter, S. Kumar, Y. Zou, R. Quintanilla, M. Sridharan, et al., Rapid and highly efficient mammalian cell engineering via Cas9 protein transfection, J. Biotechnol. 208 (2015) 44–53. Available from: https://doi.org/10.1016/j.jbiotec.2015.04.024.

[139] M.P. Zafra, E.M. Schatoff, A. Katti, M. Foronda, M. Breinig, A.Y. Schweitzer, et al., Optimized base editors enable efficient editing in cells, organoids and mice, Nat. Biotechnol. 36 (2018) 888–893. Available from: https://doi.org/10.1038/nbt.4194.

[140] J.M. Gehrke, O. Cervantes, M.K. Clement, Y. Wu, J. Zeng, D.E. Bauer, et al., An APOBEC3A-Cas9 base editor with minimized bystander and off-target activities, Nat. Biotechnol. 36 (2018) 977–982. Available from: https://doi.org/10.1038/nbt.4199.

[141] R. Kaminski, R. Bella, C. Yin, J. Otte, P. Ferrante, H.E. Gendelman, et al., Excision of HIV-1 DNA by gene editing: a proof-of-concept in vivo study, Gene Ther. 23 (2016) 690–695. Available from: https://doi.org/10.1038/gt.2016.41.

[142] X. Zhang, P. Liang, C. Ding, Z. Zhang, J. Zhou, X. Xie, et al., Efficient production of gene-modified mice using *Staphylococcus aureus* Cas9, Sci. Rep. 6 (2016). Available from: https://doi.org/10.1038/srep32565.

[143] T. Ohmori, Y. Nagao, H. Mizukami, A. Sakata, S. Muramatsu, K. Ozawa, et al., CRISPR/Cas9-mediated genome editing via postnatal administration of AAV vector cures haemophilia B mice, Sci. Rep. 7 (2017). Available from: https://doi.org/10.1038/s41598-017-04625-5.

[144] B.L. Duchêne, K. Cherif, J.-P. Iyombe-Engembe, A. Guyon, J. Rousseau, D.L. Ouellet, et al., CRISPR-induced deletion with SaCas9 restores dystrophin expression in dystrophic models in vitro and in vivo, Mol. Ther. 26 (2018) 2604–2616. Available from: https://doi.org/10.1016/j.ymthe.2018.08.010.

[145] M.E. Refaey, L. Xu, Y. Gao, B. Canan, T.M.A. Adesanya, S.C. Warner, et al., In vivo genome editing restores dystrophin expression and cardiac function in dystrophic mice, Circ. Res. 121 (2017) 923–929. Available from: https://doi.org/10.1161/CIRCRESAHA.117.310996.

[146] C. Yin, T. Zhang, X. Qu, Y. Zhang, R. Putatunda, X. Xiao, et al., In vivo excision of HIV-1 provirus by saCas9 and multiplex single-guide RNAs in animal models, Mol. Ther. 25 (2017) 1168–1186. Available from: https://doi.org/10.1016/j.ymthe.2017.03.012.

[147] Q. Wang, S. Chen, Q. Xiao, Z. Liu, S. Liu, P. Hou, et al., Genome modification of CXCR4 by *Staphylococcus aureus* Cas9 renders cells resistance to HIV-1 infection, Retrovirology 14 (2017). Available from: https://doi.org/10.1186/s12977-017-0375-0.

[148] L. Ye, J. Wang, Y. Tan, A.I. Beyer, F. Xie, M.O. Muench, et al., Genome editing using CRISPR-Cas9 to create the HPFH genotype in HSPCs: an approach for treating sickle cell disease and β-thalassemia, Proc. Natl Acad. Sci. U.S.A. 113 (2016) 10661–10665. Available from: https://doi.org/10.1073/pnas.1612075113.

[149] G. Ibba, C. Piu, E. Uleri, C. Serra, A. Dolei, Disruption by SaCas9 endonuclease of HERV-Kenv, a retroviral gene with oncogenic and neuropathogenic potential, inhibits molecules involved in cancer and amyotrophic lateral sclerosis, Viruses 10 (2018). Available from: https://doi.org/10.3390/v10080412.

[150] H. Xie, L. Tang, X. He, X. Liu, C. Zhou, J. Liu, et al., SaCas9 requires 5′-NNGRRT-3′ PAM for sufficient cleavage and possesses higher cleavage activity than SpCas9 or FnCpf1 in human cells, Biotechnol. J. 13 (2018) 1700561. Available from: https://doi.org/10.1002/biot.201700561.

[151] R. Qin, J. Li, H. Li, Y. Zhang, X. Liu, Y. Miao, et al., Developing a highly efficient and wildly adaptive CRISPR-SaCas9 toolset for plant genome editing, Plant. Biotechnol. J. 17 (2019) 706–708. Available from: https://doi.org/10.1111/pbi.13047.

[152] H. Nishimasu, L. Cong, W.X. Yan, F.A. Ran, B. Zetsche, Y. Li, et al., Crystal structure of *Staphylococcus aureus* Cas9, Cell 162 (2015) 1113–1126. Available from: https://doi.org/10.1016/j.cell.2015.08.007.

[153] B. Chen, J. Hu, R. Almeida, H. Liu, S. Balakrishnan, C. Covill-Cooke, et al., Expanding the CRISPR imaging toolset with *Staphylococcus aureus* Cas9 for simultaneous imaging of multiple genomic loci, Nucleic Acids Res. 44 (2016). Available from: https://doi.org/10.1093/nar/gkv1533. e75–e75.

[154] S.C. Strutt, R.M. Torrez, E. Kaya, O.A. Negrete, J.A. Doudna, RNA-dependent RNA targeting by CRISPR-Cas9, eLife 7 (2018). Available from: https://doi.org/10.7554/eLife.32724.

[155] R. Barrangou, C. Fremaux, H. Deveau, M. Richards, P. Boyaval, S. Moineau, et al., CRISPR provides acquired resistance against viruses in prokaryotes, Science 315 (2007) 1709–1712. Available from: https://doi.org/10.1126/science.1138140.

[156] T. Karvelis, G. Gasiunas, J. Young, G. Bigelyte, A. Silanskas, M. Cigan, et al., Rapid characterization of CRISPR-Cas9 protospacer adjacent motif sequence elements, Genome Biol. 16 (2015) 253. Available from: https://doi.org/10.1186/s13059-015-0818-7.

[157] M. Hao, Y. Cui, X. Qu, Analysis of CRISPR-Cas system in *Streptococcus thermophilus* and its application, Front. Microbiol. 9 (2018). Available from: https://doi.org/10.3389/fmicb.2018.00257.

[158] K. Xu, C. Ren, Z. Liu, T. Zhang, T. Zhang, D. Li, et al., Efficient genome engineering in eukaryotes using Cas9 from *Streptococcus thermophilus*, Cell. Mol. Life Sci. 72 (2015) 383–399. Available from: https://doi.org/10.1007/s00018-014-1679-z.

[159] W. Fujii, S. Kakuta, S. Yoshioka, S. Kyuwa, K. Sugiura, K. Naito, Zygote-mediated generation of genome-modified mice using *Streptococcus thermophilus* 1-derived CRISPR/Cas system, Biochem. Biophys. Res. Commun. 477 (2016) 473–476. Available from: https://doi.org/10.1016/j.bbrc.2016.06.070.

[160] J. Steinert, S. Schiml, F. Fauser, H. Puchta, Highly efficient heritable plant genome engineering using Cas9 orthologues from *Streptococcus thermophilus* and *Staphylococcus aureus*, Plant. J. 84 (2015) 1295–1305. Available from: https://doi.org/10.1111/tpj.13078.

[161] J. Carte, R.T. Christopher, J.T. Smith, S. Olson, R. Barrangou, S. Moineau, et al., The three major types of CRISPR-Cas systems function independently in CRISPR RNA biogenesis in *Streptococcus thermophilus*, Mol. Microbiol. 93 (2014) 98–112. Available from: https://doi.org/10.1111/mmi.12644.

[162] Y. Zhang, N. Heidrich, B.J. Ampattu, C.W. Gunderson, H.S. Seifert, C. Schoen, et al., Processing-independent CRISPR RNAs limit natural transformation in *Neisseria meningitidis*, Mol. Cell 50 (2013) 488–503. Available from: https://doi.org/10.1016/j.molcel.2013.05.001.

[163] C.M. Lee, T.J. Cradick, G. Bao, The *Neisseria meningitidis* CRISPR-Cas9 system enables specific genome editing in mammalian cells, Mol. Ther. 24 (2016) 645–654. Available from: https://doi.org/10.1038/mt.2016.8.

[164] R. Ibraheim, C.-Q. Song, A. Mir, N. Amrani, W. Xue, E.J. Sontheimer, All-in-one adeno-associated virus delivery and genome editing by *Neisseria meningitidis* Cas9 in vivo, Genome Biol. 19 (2018). Available from: https://doi.org/10.1186/s13059-018-1515-0.

[165] N. Amrani, X.D. Gao, P. Liu, A. Edraki, A. Mir, R. Ibraheim, et al., NmeCas9 is an intrinsically high-fidelity genome-editing platform, Genome Biol. 19 (2018). Available from: https://doi.org/10.1186/s13059-018-1591-1.

[166] A. Pawluk, N. Amrani, Y. Zhang, B. Garcia, Y. Hidalgo-Reyes, J. Lee, et al., Naturally occurring off-switches for CRISPR-Cas9, Cell 167 (2016) 1829–1838. Available from: https://doi.org/10.1016/j.cell.2016.11.017. e9.

[167] H. Ma, A. Naseri, P. Reyes-Gutierrez, S.A. Wolfe, S. Zhang, T. Pederson, Multicolor CRISPR labeling of chromosomal loci in human cells, Proc. Natl Acad. Sci. U.S.A. 112 (2015) 3002–3007. Available from: https://doi.org/10.1073/pnas.1420024112.

[168] Y. Zhang, R. Rajan, H.S. Seifert, A. Mondragón, E.J. Sontheimer, DNase H Activity of *Neisseria meningitidis* Cas9, Mol. Cell 60 (2015) 242–255. Available from: https://doi.org/10.1016/j.molcel.2015.09.020.

[169] B.A. Rousseau, Z. Hou, M.J. Gramelspacher, Y. Zhang, Programmable RNA cleavage and recognition by a natural CRISPR-Cas9 system from *Neisseria meningitidis*, Mol. Cell 69 (2018) 906–914. Available from: https://doi.org/10.1016/j.molcel.2018.01.025. e4.

[170] T.R. Sampson, S.D. Saroj, A.C. Llewellyn, Y.-L. Tzeng, D.S. Weiss, A CRISPR-Cas system mediates bacterial innate immune evasion and virulence, Nature 497 (2013) 254–257. Available from: https://doi.org/10.1038/nature12048.

[171] A.A. Price, T.R. Sampson, H.K. Ratner, A. Grakoui, D.S. Weiss, Cas9-mediated targeting of viral RNA in eukaryotic cells, Proc. Natl Acad. Sci. U.S.A. 112 (2015) 6164–6169. Available from: https://doi.org/10.1073/pnas.1422340112.

[172] D.H. Jo, T. Koo, C.S. Cho, J.H. Kim, J.-S. Kim, J.H. Kim, Long-term effects of in vivo genome editing in the mouse retina using *Campylobacter jejuni* Cas9 expressed via adeno-associated virus, Mol. Ther. 27 (2019) 130–136. Available from: https://doi.org/10.1016/j.ymthe.2018.10.009.

[173] T. Koo, N.B. Lu-Nguyen, A. Malerba, E. Kim, D. Kim, O. Cappellari, et al., Functional rescue of dystrophin deficiency in mice caused by frameshift mutations using *Campylobacter jejuni* Cas9, Mol. Ther. 26 (2018) 1529–1538. Available from: https://doi.org/10.1016/j.ymthe.2018.03.018.

[174] G. Dugar, R.T. Leenay, S.K. Eisenbart, T. Bischler, B. U. Aul, C.L. Beisel, et al., CRISPR RNA-dependent binding and cleavage of endogenous RNAs by the *Campylobacter jejuni* Cas9, Mol. Cell 69 (2018)

893–905. Available from: https://doi.org/10.1016/j.molcel.2018.01.032. e7.

[175] T. Yamano, B. Zetsche, R. Ishitani, F. Zhang, H. Nishimasu, O. Nureki, Structural basis for the canonical and non-canonical PAM recognition by CRISPR-Cpf1, Mol. Cell. 67 (2017) 633–645. Available from: https://doi.org/10.1016/j.molcel.2017.06.035. e3.

[176] D. Dong, K. Ren, X. Qiu, J. Zheng, M. Guo, X. Guan, et al., The crystal structure of Cpf1 in complex with CRISPR RNA, Nature 532 (2016) 522–526. Available from: https://doi.org/10.1038/nature17944.

[177] I. Fonfara, H. Richter, M. Bratovič, A. Le Rhun, E. Charpentier, The CRISPR-associated DNA-cleaving enzyme Cpf1 also processes precursor CRISPR RNA, Nature 532 (2016) 517–521. Available from: https://doi.org/10.1038/nature17945.

[178] R.E. Haurwitz, M. Jinek, B. Wiedenheft, K. Zhou, J.A. Doudna, Sequence- and structure-specific RNA processing by a CRISPR endonuclease, Science 329 (2010) 1355–1358. Available from: https://doi.org/10.1126/science.1192272.

[179] G. Zhong, H. Wang, Y. Li, M.H. Tran, M. Farzan, Cpf1 proteins excise CRISPR RNAs from mRNA transcripts in mammalian cells, Nat. Chem. Biol. 13 (2017) 839–841. Available from: https://doi.org/10.1038/nchembio.2410.

[180] L. Liu, X. Li, J. Wang, M. Wang, P. Chen, M. Yin, et al., Two distant catalytic sites are responsible for C2c2 RNase activities, Cell 168 (2017) 121–134. Available from: https://doi.org/10.1016/j.cell.2016.12.031. e12.

[181] J. Kweon, A.-H. Jang, D. Kim, J.W. Yang, M. Yoon, H. Rim Shin, et al., Fusion guide RNAs for orthogonal gene manipulation with Cas9 and Cpf1, Nat. Commun. 8 (2017). Available from: https://doi.org/10.1038/s41467-017-01650-w.

[182] R.T. Leenay, K.R. Maksimchuk, R.A. Slotkowski, R. N. Agrawal, A.A. Gomaa, A.E. Briner, et al., Identifying and visualizing functional PAM diversity across CRISPR-Cas systems, Mol. Cell 62 (2016) 137–147. Available from: https://doi.org/10.1016/j.molcel.2016.02.031.

[183] Y. Jiang, F. Qian, J. Yang, Y. Liu, F. Dong, C. Xu, et al., CRISPR-Cpf1 assisted genome editing of *Corynebacterium glutamicum*, Nat. Commun. 8 (2017) 15179. Available from: https://doi.org/10.1038/ncomms15179.

[184] J. Ungerer, H.B. Pakrasi, Cpf1 is a versatile tool for CRISPR genome editing across diverse species of Cyanobacteria, Sci. Rep. 6 (2016). Available from: https://doi.org/10.1038/srep39681.

[185] Y. Kim, S.-A. Cheong, J.G. Lee, S.-W. Lee, M.S. Lee, I.-J. Baek, et al., Generation of knockout mice by Cpf1-mediated gene targeting, Nat. Biotechnol. 34 (2016) 808–810. Available from: https://doi.org/10.1038/nbt.3614.

[186] Y. Zhang, C. Long, H. Li, J.R. McAnally, K.K. Baskin, J.M. Shelton, et al., CRISPR-Cpf1 correction of muscular dystrophy mutations in human cardiomyocytes and mice, Sci. Adv. 3 (2017). Available from: https://doi.org/10.1126/sciadv.1602814.

[187] M. Yang, H. Wei, Y. Wang, J. Deng, Y. Tang, L. Zhou, et al., Targeted disruption of V600E-mutant BRAF gene by CRISPR-Cpf1, Mol. Ther. Nucleic Acids 8 (2017) 450–458. Available from: https://doi.org/10.1016/j.omtn.2017.05.009.

[188] D. Kim, J. Kim, J.K. Hur, K.W. Been, S. Yoon, J.-S. Kim, Genome-wide analysis reveals specificities of Cpf1 endonucleases in human cells, Nat. Biotechnol. 34 (2016) 863–868. Available from: https://doi.org/10.1038/nbt.3609.

[189] D. Kim, S. Kim, S. Kim, J. Park, J.-S. Kim, Genome-wide target specificities of CRISPR-Cas9 nucleases revealed by multiplex Digenome-seq, Genome Res. 26 (2016) 406–415. Available from: https://doi.org/10.1101/gr.199588.115.

[190] E. Tóth, N. Weinhardt, P. Bencsura, K. Huszár, P.I. Kulcsár, A. Tálas, et al., Cpf1 nucleases demonstrate robust activity to induce DNA modification by exploiting homology directed repair pathways in mammalian cells, Biol. Direct. 11 (2016). Available from: https://doi.org/10.1186/s13062-016-0147-0.

[191] Y. Wang, K.I. Liu, N.-A.B. Sutrisnoh, H. Srinivasan, J. Zhang, J. Li, et al., Systematic evaluation of CRISPR-Cas systems reveals design principles for genome editing in human cells, Genome Biol. 19 (2018). Available from: https://doi.org/10.1186/s13059-018-1445-x.

[192] B. Li, W. Zhao, X. Luo, X. Zhang, C. Li, C. Zeng, et al., Engineering CRISPR-Cpf1 crRNAs and mRNAs to maximize genome editing efficiency, Nat. Biomed. Eng. 1 (2017). Available from: https://doi.org/10.1038/s41551-017-0066.

[193] X. Zhang, J. Wang, Q. Cheng, X. Zheng, G. Zhao, J. Wang, Multiplex gene regulation by CRISPR-ddCpf1, Cell Discov. 3 (2017) 17018. Available from: https://doi.org/10.1038/celldisc.2017.18.

[194] B.P. Kleinstiver, A.A. Sousa, R.T. Walton, Y.E. Tak, J.Y. Hsu, K. Clement, et al., Engineered CRISPR–Cas12a variants with increased activities and improved targeting ranges for gene, epigenetic and base editing, Nat. Biotechnol. 37 (2019) 276–282. Available from: https://doi.org/10.1038/s41587-018-0011-0.

[195] X. Zhang, W. Wang, L. Shan, L. Han, S. Ma, Y. Zhang, et al., Gene activation in human cells using CRISPR/Cpf1-p300 and CRISPR/Cpf1-SunTag systems, Protein Cell 9 (2018) 380–383. Available from: https://doi.org/10.1007/s13238-017-0491-6.

[196] O.O. Abudayyeh, J.S. Gootenberg, P. Essletzbichler, S. Han, J. Joung, J.J. Belanto, et al., RNA targeting with CRISPR-Cas13a, Nature 550 (2017) 280–284. Available from: https://doi.org/10.1038/nature24049.

[197] D.B.T. Cox, J.S. Gootenberg, O.O. Abudayyeh, B. Franklin, M.J. Kellner, J. Joung, et al., RNA editing with CRISPR-Cas13, Science 358 (6366) (2017) 1019–1027.

[198] J.S. Gootenberg, O.O. Abudayyeh, J.W. Lee, P. Essletzbichler, A.J. Dy, J. Joung, et al., Nucleic acid detection with CRISPR-Cas13a/C2c2, Science 356 (2017) 438–442. Available from: https://doi.org/10.1126/science.aam9321.

CHAPTER

3

Genome engineering with CRISPR/Cas9, ZFNs, and TALENs

Angela Y. Chang

Edward S. Harkness Eye Institute, Department of Ophthalmology, Columbia University Irving Medical Center, New York, NY, United States

Introduction

New technologies for genetic engineering including zinc-finger nucleases (ZFN), transcription activator-like effector nucleases (TALENs), and clustered regularly interspaced short palindromic repeat (CRISPR)/Cas9-guided endonucleases allow site-specific genomic editing and have emerged as promising therapeutic strategies for treating different disorders in vivo. All three technologies facilitate genome editing by inducing a double-strand break (DSB) in the DNA at the target site. Each system uses homologous binding proteins or RNA to mediate specificity of the DSBs in the target DNA (Fig. 3.1). Following the introduction of a DBS in DNA, repair occurs through one of the two pathways: the error-prone nonhomologous end joining (NHEJ) pathway or the high-fidelity homology-directed repair (HDR) pathway.

The NHEJ pathway is the default pathway and occurs in the absence of a repair template. This error-prone pathway frequently results in a large scale of insertion/deletion (indel) mutations causing frameshift or nonsense mutations that lead to impaired gene expression or gene knockouts. In the presence of a repair template, the DSB can be repaired through the HDR pathway. This DSB can subsequently be used to introduce specific exogenous mutations or modifications at the site of the DSB. Each system is unique in how it recognizes target DNA sites and generates site-specific mutations; they even have different requirements for specificity and target DNA characteristics. While each technique has its benefits, CRISPR/Cas9 offers significant advantages over both ZFN and TALEN technologies due to its ease of design and delivery, as well as its ability to introduce different mutations simultaneously. The three techniques also differ in their specificity and efficiency, with TALEN and CRISPR/Cas9 offering advantages over ZFN. In this chapter, we will review all three genetic engineering techniques, describing and comparing their function and efficacy as well as discussing their applications.

CRISPR Genome Surgery in Stem Cells and Disease Tissues
DOI: https://doi.org/10.1016/B978-0-12-817876-8.00009-7

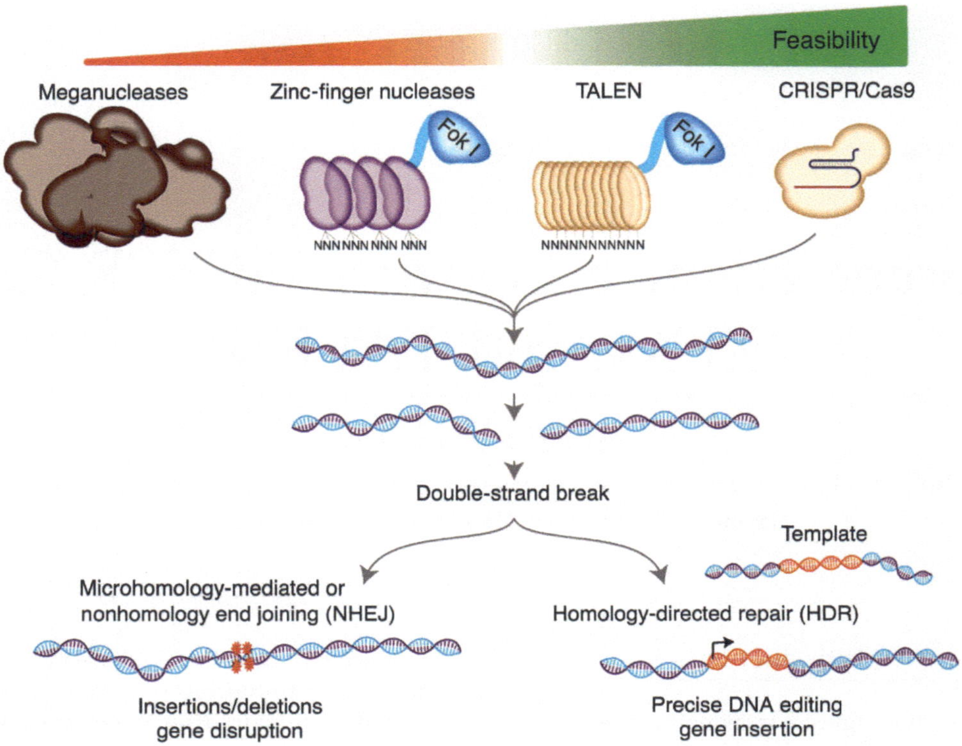

FIGURE 3.1 Zinc-finger nucleases, transcription activator-like effector nucleases (TALENs), and CRISPR (clustered regularly interspaced short palindromic repeat)/Cas9 technologies can all be used to induce a double-strand break in DNA. Subsequent repair occurs via the error-prone nonhomology end joining pathway or the high-fidelity homology-directed repair pathway if template DNA is present. (File licensed under the Creative Commons Attribution-Share Alike 4.0 International license.)

Zinc-finger nucleases

The Cys2–His2 zinc-finger domain is the most common DNA-binding motif in eukaryotes [1]. A zinc ion coordinates cysteine and histidine, leading to two beta sheets and an alpha helix. Over 30 types of zinc-finger motifs have been identified, most have different combinations of cysteine and histidine, including Cys2–His, Cys2–Cys–His, and Cys2–Cys2 [2]. Other types of zinc-finger domains are RING domain proteins found in ubiquitin ligases, the PHD domain involved in epigenetic modifications, and the LIM domain found in transcription factors important for cytoskeleton interaction [2].

ZFNs are fusion proteins composed of two domains: a zinc-finger protein (ZFP) domain on the *N*-terminus that binds DNA and mediates specificity, and a nuclease domain, typically restriction endonuclease Fok1, on the *C*-terminus that cleaves DNA to induce a DBS. Each zinc finger consists of approximately 30 amino acids in a $\beta\beta\alpha$ arrangement [1]. This configuration is stabilized through hydrophobic interactions and the coordination of a zinc ion with the cysteine and histidine residues [1]. Each ZFP recognizes a 3-base pair (bp) DNA sequence, mediating specificity. Three to six ZFPs can be modulated to construct a zinc-finger domain that can target unique 9–18 nucleotide DNA sequences [1].

The nuclease domain consists of the catalytic domain of the restriction enzyme Fok1. This nuclease domain can be combined with ZFPs to form ZFNs, thereby generating site-specific genetic editing. Fok1 nuclease activity necessitates heterodimerization of two unique ZFN monomers, left and right. The left and right ZFN monomers bind to opposite strands of the DNA separated by 5–7 bp. Upon binding of the ZFN monomers, the respective Fok1 domains dimerize and generate a DSB, leaving behind 5′ overhangs [3]. Subsequent repair can occur through the error-prone NHEJ pathway, that is disposed to indel mutations that lead to frameshift mutations. This shift in the reading frame can cause a knockout of the gene.

ZFN-guided genomic editing has been utilized in this manner to create knockouts of specific genes. One group utilized ZFN to target the first coding exon of human C–C motif chemokine receptor 5 (CCR5), a chemoreceptor found on white blood cells [4]. It is also a target the human immunodeficiency virus (HIV)-1 uses to enter target cells. Use of ZFNs were introduced to human cell lines, leading to reduced CCR5 expression. Use of this CCR5 ZFN produced a protective effect against HIV-1 in mouse models [4].

Repair after the DSB induced by ZFN can also occur via the higher fidelity HDR pathway. In this manner, ZFN has been used to correct gene mutations in diseases such as hemophilia B, characterized by deficient factor IX [5]. By introducing the wild-type sequence, subsequent gene expression yielded functional copies of the wild-type protein. In mouse models of hemophilia B, this method of gene delivery led to normal coagulation function, lasting for more than 30 weeks [5].

A major limitation of using ZFN is that even with successful dimerization of monomers, many fail to cleave chromosomal DNA, and exhibit varying levels of binding affinity and activity [3,6]. Target sites that are rich in guanine nucleotides and consist of 5′-GNN-3′ have been found to be more successful [6]. Additionally, engineering ZFNs with open-source libraries limits the number of target sites and may take significant time to optimize [3]. Laboratory-made ZFNs may also be cytotoxic to cells, in contrast to TALEN and CRISPR/Cas9 systems [6]. Additionally, ZFNs display off-target activity through homodimerization of ZFN monomers and at genomic regions with sequence similarity to the target site. One method for increasing specificity is to use ZFN with distinct Fok1 domains that are obligate heterodimers [7].

Transcription activator-like effector nucleases

Transcription activator-like effectors (TALEs) are DNA-binding proteins found in Gram-negative plant pathogen *Xanthomonas* that modulate gene transcription in plants during pathogenesis [8,9]. The DNA-binding domain consists of 34-amino acid tandem repeats flanked by nonrepetitive sequences in the N and C termini. The number of repeats can range from 1.5 to 33.5, ending with a half repeat sequence. Each of these repeat sequences is identical except for the 12th and 13th amino acid sequences, which mediate nucleotide-binding specificity. These positions, termed the repeat variable diresidues (RVDs), correspond to different nucleotide bases, where NI = A, HD = C, NG = T, and NN = G or A [9]. Thus each DNA-binding domain can recognize a specific nucleotide, and different TALE modules can be linked together to recognize a specific DNA sequence. They also include a domain that activates gene transcription. Thus TALEs with unique DNA recognition sites can be constructed with high sequence specificity by changing the number and type of TALE modules. TALE modules linked to the endonuclease domain of Fok1, termed TALENs, can be used for site-specific gene editing. Much like ZFNs, TALENs require heterodimerization of two TALEN monomers to form the active Fok1 endonuclease that can generate DSBs.

As previously discussed with ZFNs, DSBs induced by TALENs can be repaired using NHEJ or HDR. TALENs have been used as gene therapy in studies, without a repair template to induce knockouts and with a repair template to restore protein function. One study utilized TALEN gene editing to restore dystrophin, the mutated protein in Duchenne muscular dystrophy (DMD) [10]. Most of the mutations seen in DMD are frameshift mutations from deletions. This study utilized editing at exon 51 of the dystrophin gene to correct the frameshift mutation via the NHEJ pathway [10]. Using this method of gene therapy, there is lasting presence of the restored dystrophin gene that can also be passed to cell progeny [10].

TALEN gene editing has also been used to edit genes including NTF3 and CCR5, genes implicated in HIV and neurodegenerative disease, respectively [11]. This method allowed for modification of gene expression using both NHEJ and HDR-directed repair. This method of gene editing has also been used in immunotherapy to efficiently manufacture autologous T cells with chimeric antigen receptors. These autologous cells do not self-react and lack T-cell receptor and CD52 components, targets of chemotherapeutic agents [12].

Compared to ZFN and CRISPR/Cas9, TALENs have fewer constraints when it comes to choosing a given target DNA sequence, giving this method a unique advantage. In contrast to ZFN modules, which target 3-nucleotide sequences, TALE modules are able to target single nucleotides, leading to fewer constraints on target site selection, compared to ZFN modules that require 3-nucleotide sequences or the single-guide RNA (sgRNA) sequence in CRISPR/Cas9. Additionally, the TALE repeat array can be extended to any length, unlike in ZFNs, which are limited to 9–18 bp sequences. However, longer TALE repeat domains may lead to increased off-target activity since they are better able to tolerate base pair mismatches [13]. As with ZFNs, the use of monomers that are obligate heterodimers is one method of decreasing off-target activity [3,13]. A limitation to using TALEs is the requirement of a thymine on the 5′ end of the target DNA sequence [6]; this ultimately decreases the number of potential target sites. This thymine is recognized by a cryptic repeat structure that is involved in DNA binding [1]. Although TALEs cannot be used to cleave target DNA with methylated cytosine nucleotides, using the RVD that specifies thymine nucleotides has been shown to cleave DNA with methylated cytosine residues [6].

Both ZFNs and TALENs are dependent on Fok1 nuclease activity, but the types of mutations favorable for each system varies: TALENs are more favorable for producing deletions rather than insertions, while ZFNs can induce both insertions and deletions [14]. A major disadvantage of using TALENs compared to ZFNs, however, is their larger size, which may impair their ability to be delivered therapeutically using viral vectors of limited cargo size [3]. Experimentally, the cDNA coding TALENs can be three times larger than that for ZFNs, creating additional hurdles for delivery and expression in cells [3]. TALENs can also contain up to 20 TALE modules, that contain highly homologous sequences, and result in DNA recombination that changes sequence specificity or impairing its delivery using viral vectors [3].

CRIPSR/Cas9

The CRISPR/Cas9 system is a microbial-derived adaptive immune system that can be used for targeted genomic engineering. The CRISPR/Cas9 system recognizes its target through protospacer adjacent motifs (PAMs) flanking the target site, complementary base pairing between RNA and DNA, and interactions between Cas9 and target DNA. Cas9 is an endonuclease guided by variable CRISPR RNA sequences (crRNA) and invariable trans-activating RNA sequences. These RNA sequences complex with Cas9 to form the active endonuclease complex capable of site-specific DNA mutagenesis. The crRNA sequence mediates specificity and is complementary to the target DNA through Watson–Crick base pairing. The PAM sequence is a 2–5 bp motif necessary for Cas nuclease recognition and cleavage of target DNA. Cas9 induces a DSB in the DNA 3 bp upstream of the PAM sequence. In the CRISPR type II system of *Streptococcus pyogenes*, the PAM sequence is 5′-NGG-3′, with N being any nucleotide. While potential target sites are limited by the presence of PAM sequence, in this system, since the NGG PAM sequence is found every 8–12 nucleotides in the human genome [15,16], the number of potential target sites is not severely limited [6]. However, not all potential targets are cleaved at the same rates with the same efficiency [6]. Different sgRNA sequences targeting the same gene can also have efficiencies that vary up to 10-fold depending on off-target effects, concentrations of sgRNA, and GC content of the nucleotides proximal to the PAM sequences [17].

CRISPR/Cas9 been used successfully as a method for mutagenesis in mammalian cells, zebrafish, bacterial cells, nematodes, fruit flies, and human pluripotent stem cells [6]. One study used this method of genetic engineering to disrupt expression of PCSK9, a protein involved in regulating low-density lipoprotein receptor levels and blood cholesterol, in mouse models. These mice demonstrated a 35%–40% decrease in cholesterol levels compared to controls [18]. A similar approach could be used to therapeutically decrease cholesterol levels in humans.

Another study used CRISPR/Cas9 and HDR repair to target the Crygc gene; an autosomal dominant mutation of this gene causes cataracts [19]. Cas9 and sgRNA coding for the wild-type protein were injected into mouse zygotes, leading to expression of the wild-type protein and correction of the mutation [19]. The resulting mice were able to transmit the corrected allele to progeny [19]. This study, among others, demonstrates the therapeutic potential of using CRISPR/Cas9 to correct genetic mutations.

A major advantage to using CRISPR/Cas9 is ease of design and preparation; crRNA and sgRNA can easily be cloned with plasmids and protein engineering is not required [6]. Essentially, any 20-nucleotide sequence can be targeted by engineering the sgRNA. The Cas9 protein does not require any protein engineering since the enzymatic machinery remains constant regardless of application. Furthermore, Cas9 induces the DSB without need for dimerization, unlike FokI in ZFN and TALENs [8].

The biggest drawback to using the CRISPR/Cas9 system is its potential for off-target effects. Factors mediating sgRNA target specificity include the amount and distribution of base pair mismatches between sgRNA and target DNA [17]. Since sgRNA is only 19–21 bp and up to 5 bp mismatches between the sgRNA and DNA can be tolerated, there can be numerous potential sites for off-target cleavage [20]. The presence of either an extra base on the DNA (DNA bulge) or a missing base (sgRNA bulge) when base pairing with the sgRNA variable region can lead to off-target CRISPR/Cas9 activity. Specifically, sgRNA bulges of up to 4 nucleotides can be tolerated, leading to off-target activity [20]. As ZFN and TALEN systems require heterodimerization of monomers at a specific site, they may confer more specificity than the CRISPR/Cas9 system, which relies on a single RNA-guided endonuclease. One method to increase specificity and decrease off-target activity involves using a paired nickase strategy that generates two single-strand breaks.

This method involves using two different sgRNA sequences guiding a mutant version of the Cas9 protein to the target site, increasing the requirements for specificity [15,21].

The CRISPR/Cas9 system offers a significant advantage over ZFNs and TALENs largely due to its simpler design and preparation [3]. The system is able to achieve site-specific mutagenesis with a 20 bp variable crRNA sequence, which can be incorporated into an RNA plasmid backbone. ZFNs and TALENs require more significant protein engineering, while the Cas9 endonuclease remains constant. Additionally, CRISPR/Cas9-mediated gene editing allows for simultaneous mutation of multiple genes with the addition of multiple sgRNAs [3].

However, all three systems have unique advantages and disadvantages over the others in their requirements mediating specificity, success rate, off-target effects, and mutation rate. A summary of these differences can be found in Table 3.1. Specifically, the success rate in using TALENs (99%) and CRISPR/Cas9 (90%) offers significant advantages over ZFN (24%) [6]. Use of ZFN is also associated with high off-target effects compared to TALENs, while CRISPR/Cas9 may have variable off-target effects depending on the sgRNA sequence. The target of genetic engineering should also be considered, as each of these systems have different target site requirements. ZFN perform best with G-rich target sequences, TALENs require the sequence to start with T at the 5′ end of the target sequence, and CRISPR/Cas9 requires presence of a PAM sequence [6].

TABLE 3.1 A comparison of using ZFN, TALENs, and CRISPR/Cas9 in gene editing [6].

	ZFNs	TALENs	RGENs
DNA targeting specificity determinant	Zinc-finger proteins	Transcription activator-like effectors	crRNA or sgRNA
Nuclease	FokI	FokI	Cas9
Success rate[a]	Low (~24%)	High (>99%)	High (~90%)
Average mutation rate[b]	Low or variable (~10%)	High (~20%)	High (~20%)
Specificity-determining length of target site	18–36 bp	30–40 bp	22 bp (total length 23 bp)
Restriction in target site	G-rich	Start with T and end with A (owing to the heterodimer structure)	End with an NGG or NAG (lower activity) sequence (i.e., PAM)
Design density	One per ~100 bp	At least one per base pair	One per 8 bp (NGG PAM) or 4 bp (NGG and NAG PAM)
Off-target effects	High	Low	Variable
Cytotoxicity	Variable to high	Low	Low
Size	~1 kb × 2	~3 kb × 2	4.2 kb (Cas9 from *Streptococcus pyogenes*) + 0.1 kb (sgRNA)

[a]*The success rate is defined as the proportion of nucleases that induce mutations at frequencies >0.5% in HEK293 cells.*
[b]*The average mutation rate is based on frequency of nonhomologous end-joining-mediated insertions and deletions obtained at the nuclease target site.*
Notes: CRISPR, clustered regularly interspaced short palindromic repeat; *crRNA*, CRISPR RNA sequences; *PAM*, protospacer adjacent motif; *sgRNA*, single-guide RNA; *TALENs*, transcription activator-like effector nucleases; *RGENs*, CRISPR/Cas-based RNA-guided DNA endonucleases; *ZFNs*, zinc-finger nucleases.

Conclusion

Genomic editing technologies including ZFNs, TALENs, and CRIPSR/Cas9 generate DSBs in DNA, allowing for precise mutagenesis through the HDR pathway. Already these technologies have been key in studies examining therapeutic strategies in a wide range of diseases including tyrosinemia [22], cystic fibrosis [23], X-linked severe combined immune deficiency [24], hemophilia [5], and sickle cell disease [25]. Further optimization to improve efficiency and specificities, which currently varies between these systems, is needed to enhance their potentials to be used in clinical applications.

References

[1] T. Gaj, C.A. Gersbach, C.F. Barbas 3rd, ZFN, TALEN, and CRISPR/Cas-based methods for genome engineering, Trends Biotechnol. 31 (7) (2013) 397–405.

[2] M. Cassandri, et al., Zinc-finger proteins in health and disease, Cell Death Discov. 3 (2017) 17071.

[3] R.M. Gupta, K. Musunuru, Expanding the genetic editing tool kit: ZFNs, TALENs, and CRISPR-Cas9, J. Clin. Invest. 124 (10) (2014) 4154–4161.

[4] E.E. Perez, et al., Establishment of HIV-1 resistance in CD4+ T cells by genome editing using zinc-finger nucleases, Nat. Biotechnol. 26 (7) (2008) 808–816.

[5] H. Li, et al., In vivo genome editing restores haemostasis in a mouse model of haemophilia, Nature 475 (7355) (2011) 217–221.

[6] H. Kim, J.S. Kim, A guide to genome engineering with programmable nucleases, Nat. Rev. Genet. 15 (5) (2014) 321–334.

[7] Y. Doyon, et al., Enhancing zinc-finger-nuclease activity with improved obligate heterodimeric architectures, Nat. Methods 8 (1) (2011) 74–79.

[8] J.S. LaFountaine, K. Fathe, H.D. Smyth, Delivery and therapeutic applications of gene editing technologies ZFNs, TALENs, and CRISPR/Cas9, Int. J. Pharm. 494 (1) (2015) 180–194.

[9] N.E. Sanjana, et al., A transcription activator-like effector toolbox for genome engineering, Nat. Protoc. 7 (1) (2012) 171–192.

[10] D.G. Ousterout, et al., Reading frame correction by targeted genome editing restores dystrophin expression in cells from Duchenne muscular dystrophy patients, Mol. Ther. 21 (9) (2013) 1718–1726.

[11] J.C. Miller, et al., A TALE nuclease architecture for efficient genome editing, Nat. Biotechnol. 29 (2) (2011) 143–148.

[12] L. Poirot, et al., Multiplex genome-edited T-cell manufacturing platform for "off-the-shelf" adoptive T-cell immunotherapies, Cancer Res. 75 (18) (2015) 3853–3864.

[13] J.P. Guilinger, et al., Broad specificity profiling of TALENs results in engineered nucleases with improved DNA-cleavage specificity, Nat. Methods 11 (4) (2014) 429–435.

[14] Q. Ul Ain, J.Y. Chung, Y.H. Kim, Current and future delivery systems for engineered nucleases: ZFN, TALEN and RGEN, J. Control. Release 205 (2015) 120–127.

[15] F.A. Ran, et al., Genome engineering using the CRISPR-Cas9 system, Nat. Protoc. 8 (11) (2013) 2281–2308.

[16] X. Wu, et al., Genome-wide binding of the CRISPR endonuclease Cas9 in mammalian cells, Nat. Biotechnol. 32 (7) (2014) 670–676.

[17] X. Ren, et al., Enhanced specificity and efficiency of the CRISPR/Cas9 system with optimized sgRNA parameters in Drosophila, Cell Rep. 9 (3) (2014) 1151–1162.

[18] Q. Ding, et al., Permanent alteration of PCSK9 with in vivo CRISPR-Cas9 genome editing, Circ. Res. 115 (5) (2014) 488–492.

[19] Y. Wu, et al., Correction of a genetic disease in mouse via use of CRISPR-Cas9, Cell Stem Cell 13 (6) (2013) 659–662.

[20] Y. Lin, et al., CRISPR/Cas9 systems have off-target activity with insertions or deletions between target DNA and guide RNA sequences, Nucleic Acids Res. 42 (11) (2014) 7473–7485.

[21] F.A. Ran, et al., Double nicking by RNA-guided CRISPR Cas9 for enhanced genome editing specificity, Cell 154 (6) (2013) 1380–1389.

[22] H. Yin, et al., Genome editing with Cas9 in adult mice corrects a disease mutation and phenotype, Nat. Biotechnol. 32 (6) (2014) 551–553.

[23] G. Schwank, et al., Functional repair of CFTR by CRISPR/Cas9 in intestinal stem cell organoids of cystic fibrosis patients, Cell Stem Cell 13 (6) (2013) 653–658.

[24] F.D. Urnov, et al., Highly efficient endogenous human gene correction using designed zinc-finger nucleases, Nature 435 (7042) (2005) 646–651.

[25] J. Zou, et al., Site-specific gene correction of a point mutation in human iPS cells derived from an adult patient with sickle cell disease, Blood 118 (17) (2011) 4599–4608.

CHAPTER

4

Single-guide RNAs: rationale and design

Angela Y. Chang

Edward S. Harkness Eye Institute, Department of Ophthalmology, Columbia University Irving Medical Center, New York, NY, United States

CRISPR/Cas9 background and the role of sgRNA

The clustered regularly interspaced short palindromic repeats (CRISPR)/Cas9 system is a microbial adaptive immune system that can be used for targeted genomic engineering with clinical therapeutic potential. Following exposure to pathogens, bacteria and archaea incorporate short fragments of foreign DNA from viruses or plasmids into their chromosomes to keep a genetic record of the pathogen. These short fragments are incorporated as new spacer elements between constant CRISPR sequences in the CRISPR loci in their genome. Transcription of the CRISPR DNA array yields CRISPR RNAs (crRNA) that interact with trans-activating RNA (tracrRNA) and Cas nucleases to mediate targeted destruction of the foreign DNA during subsequent exposures. The Cas nuclease target is mediated by the variable crRNA sequence that possesses complementarity to the target DNA site though Watson–Crick base pairing. By changing the crRNA sequence, the Cas nuclease can be programmed to target novel DNA sites, allowing for site-specific genomic editing Fig. 4.1.

Three CRISPR systems (I–III) have been characterized in a range of hosts including *Streptococcus pyogenes*, *Streptococcus thermophiles*, and *Neisseria meningiditis*. The most well-studied and commonly used system is the type II CRISPR/Cas9 system in *S. pyogenes*. Cas9 is the RNA-guided endonuclease in the type II CRISPR system. It induces a blunt ended double-stranded break (DSB) in the DNA at the site specified by the crRNA sequence. Target recognition in the CRISPR/Cas9 system requires the presence of protospacer adjacent motifs (PAMs) flanking the target site, complementary base pairing between guide RNA and DNA, and interactions between Cas9 and target DNA. The PAM sequence is a 2–6 base pair (bp) motif located near the target DNA sequence that is necessary for Cas nuclease recognition of DNA and target cleavage [1]. Cas9 induces a DSB in the DNA 3 bp upstream of the PAM sequence [1]. The PAM sequence required for CRISPR/Cas9 function is dependent on the origin of the CRISPR system. In the

CRISPR Genome Surgery in Stem Cells and Disease Tissues
DOI: https://doi.org/10.1016/B978-0-12-817876-8.00003-6

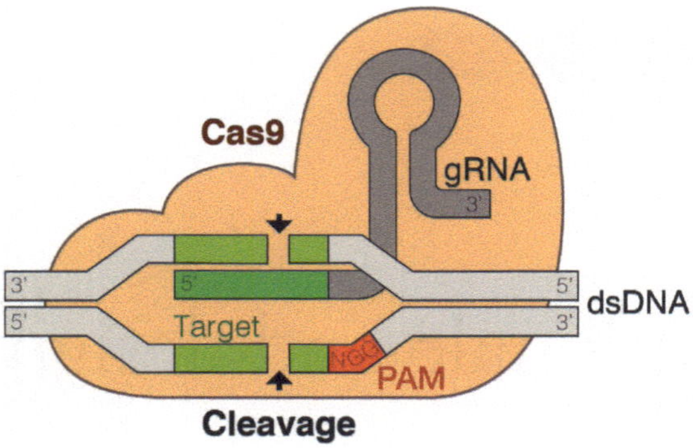

FIGURE 4.1 Cas 9 (*orange*) cleavage of dsDNA depends on sgRNA, which consists of a complex of crRNA and tracrRNA. The crRNA sequence can be programmed to target a specific DNA site, allowing for site-specific genomic editing. It is complementary to the DNA directly 5′ of the site of interest. TracrRNA is a noncoding sequence that facilitates crRNA processing, maturation, and binding to Cas9. *PAM*, Protospacer adjacent motifs. *Courtesy of Creative Commons Attribution-Share Alike 4.0 International license.*

type II system of *S. pyogenes*, the PAM sequence is 5′−NGG−3′, with N being any nucleotide. This PAM sequence must be 3′ of the target DNA sequence [1]. Thus the potential sites that are able to be targeted by the CRISPR/Cas9 system are limited by the presence of PAM sequence. However, since the NGG PAM sequence is found every 8−12 nucleotides in the human genome [1,2], the number of potential target sites is not severely limited [3].

Repair of the DNA DSB following Cas9 cleavage occurs through one of the two pathways: the error-prone nonhomologous end joining (NHEJ) pathway or the high-fidelity homology-directed repair (HDR) pathway, as shown in Fig. 4.2. The NHEJ pathway is the default pathway in eukaryotic cells and takes place in the absence of a repair template. Religation of the DNA strands occurs through

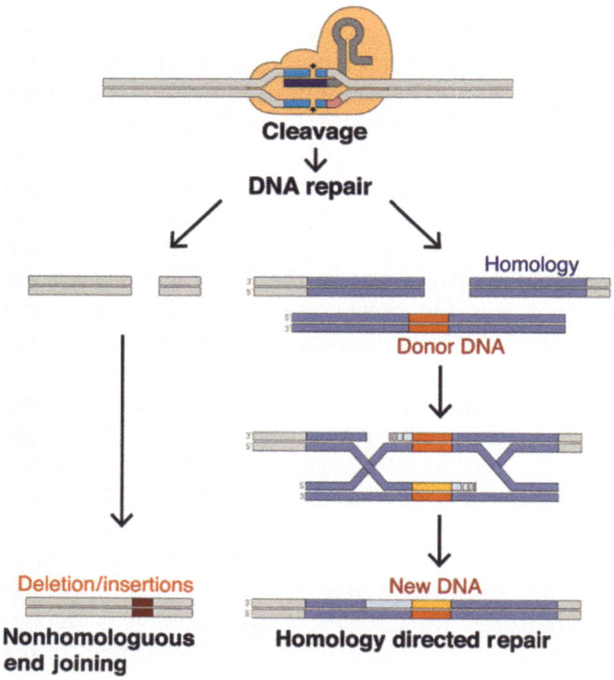

FIGURE 4.2 Repair of the double-stranded breaks in DNA following Cas9 cleavage occurs through one of the two pathways: the error-prone nonhomologous end joining (NHEJ) pathway or the high-fidelity homology-directed repair pathway. The NHEJ pathway is the default pathway in eukaryotic cells and takes place in the absence of a repair template. The NHEJ pathway is associated with indel mutations. *Courtesy of Creative Commons Attribution-Share Alike 4.0 International license.*

the activity of endogenous protein factors [4]. This error-prone pathway frequently results in insertion/deletion (indel) mutations, causing frameshift and nonsense mutations that lead to impaired gene expression or gene knockouts. In the presence of an endogenous or exogenous repair template with homology to the sites adjacent to the DSB, the DNA can be repaired through the high-fidelity HDR pathway. The HDR pathway repairs the DSB according to the template. Exogenous templates can be introduced into the CRISPR/Cas9 system in the form of DNA or single-strand oligodeoxynucleotide (ssODN) in order to introduce specific mutations and modifications. The HDR pathway, however, only occurs in actively dividing cells in the S phase of the cell cycle, and its efficiency also varies based on the site of the DSB and characteristics of the repair template [1,5].

Design of sgRNA

Single-guide RNA (sgRNA) is a synthetic complex of the native crRNA and tracrRNA necessary for site-specific CRISPR/Cas9 DSBs. The crRNA is a variable 20-nucleotide sequence that engages in Watson–Crick base pairing with the target sequence on the double-stranded DNA, regulating target specificity. The tracrRNA is a noncoding sequence with homology to the CRISPR repeat sequence that facilitates the processing and maturation of the crRNA as well as binding to Cas9. The crRNA is complementary to the DNA directly 5′ of the site of interest. The presence of the PAM sequence on the 3′ end is required for cleavage but does not count as part of the 20-nucleotide crRNA sequence. The PAM sequence can be on either the sense or antisense strand of DNA, and the variable region consists of the 20-nucleotide sequence upstream, not including the PAM sequence.

The sgRNA is also key in activating and regulating Cas9 enzymatic activity. Cas9 typically remains in an autoinhibited conformation [6], but interaction of Cas9 with crRNA and tracrRNA sequences induces Cas9 to form the active DNA endonuclease complex [3]. Two domains within Cas9, the conserved PAM recognition loops and the variable sgRNA interaction region, dictate its specificity. The interaction between sgRNA and Cas9 induces conformational changes in Cas9 to its catalytically active state [6]. Upon these interactions, Cas9 forms a central channel that is involved in the binding and cleaving of the target DNA [6]. This central channel contains the two catalytically active sites, one for each of the target DNA strands, on opposite sides of the channel DNA [3,6].

Online tools to aid in sgRNA design

Multiple sgRNA sequences can be constructed for a given target and each of these sequences can be assessed for cleavage efficiency and likelihood of off-target activity. Online tools such as the CRISPR Design Tool [1] can aid in the sgRNA design process by highlighting potential target sites for a region of interest in the genome, the presence of off-target sites in the genome, and the likelihood of off-target activity. The tools also provide information about primers that can be used to assemble sgRNA constructs.

Factors mediating sgRNA targeting specificity

Various sgRNA sequences can be constructed for engineering a specific mutation in a target gene, and their efficiencies can vary up to 10-fold [7]. Optimizing sgRNA activity is thus crucial for effective CRISPR/Cas9 function at a specified target. Minimizing off-target activity while maximizing on-target activity is two key considerations in sgRNA design [1].

Identifying off-target DNA sequences and their likelihood of interacting with the sgRNA is crucial for minimizing off-target activity. Factors affecting the likelihood of mismatch base pairing between sgRNA and DNA at off-target sites include similarity of the off-target sequence to the target sequence, whether the sequence is followed by a PAM sequence, the location and distribution of the mismatches on the sgRNA sequence, the number of mismatches between the sgRNA and DNA sequences, and the amount of sgRNA present relative to Cas9 [1,8]. Up to 5 bp mismatches between the sgRNA and DNA can be tolerated, leading to numerous potential sites for off-target cleavage [9]. The presence of either an extra base on the DNA (DNA bulge) or a missing base (sgRNA bulge) when base pairing with the sgRNA variable region can lead to off-target CRISPR/Cas9 activity [9]. In particular, sgRNA bulges of up to four nucleotides can be tolerated. Additionally, bases on the 3′ end of the sgRNA near the PAM sequence generally do not tolerate mismatches, while those on the 5′ end are more tolerable [8]. To minimize off-target activity, potential off-target regions should not be followed by a PAM sequence, contain greater than 3 bp mismatches with at least two being within the "seed region" directly proximal to the PAM sequence, and should have either a consecutive distribution of mismatches or have 3 bp mismatches within 4 consecutive nucleotides [8].

sgRNA design criteria to optimize specificity and efficiency while minimizing off-target activity

Intrinsic features of the sgRNA, specifically the nucleotide composition of the 20-nucleotide variable sequence, play a role in mediating sgRNA efficiency. sgRNA targeting sequences with very high (≥80%) or very low (≤35%) GC content are less effective in CRISPR/Cas9-mediated gene editing [10]. Sequences with very high GC content may be less effective because very high GC content can lead to better hybridization and increased stability of interactions between the sgRNA and DNA sequence, even in the presence of base pair mismatches, facilitating off-target CRISPR/Cas9 activity [9]. sgRNA sequences with very low GC content may not have enough GC base pairing interactions necessary for sgRNA stability and activity. In particular, the GC content of the 6 nucleotides upstream of the PAM sequence can be predictive of sgRNA efficiency. sgRNAs with over 50% GC content in this 6-nucleotide region have higher efficiencies, possibly due to increased hydrogen bonding between the sgRNA and the DNA target [7]. Additionally, enrichment of guanine nucleotides at the 5′ end of the sgRNA sequence may result in the formation of G-quadruplexes that are protective against exonuclease degradation [11]. Studies have also shown that guanine-rich and adenine-depleted sgRNAs have increased stability, and thus increased Cas9 activity [11].

Specifically, the bases located within the "seed" region, encompassing 10–12 nucleotides upstream of PAM, seem to host the most significant differences determining sgRNA efficiency. Some studies have shown that as few as 5 nucleotides within this "seed" region directly upstream of the PAM are needed for sgRNA and DNA binding specificity [2]. This "seed" region plays a crucial role in pairing the sgRNA with the target DNA, mediating specificity and efficiency. The seed sequence composition is important, since the frequency in which the seed sequence followed by a PAM sequence occurs in the genome dictates the number of potential off-target sites [2]. Additionally, the seed sequence also affects the amount of sgRNA present and the loading of Cas9 [2]. A two-state model of Cas9 binding and cleavage has been proposed: the complementarity in the seed region triggers Cas9 binding, and extensive base pairing with the target DNA mediates Cas9 cleavage of target DNA [2].

In the seed region, the final 4 nucleotides on the 3′ end of the crRNA proximal to the PAM sequence is the most important determinant in Cas9 loading, that is, sgRNAs with purines in the last 4 nucleotides are favored while those with pyrimidines, especially thymine (T), are disfavored [10]. Furthermore, the final base directly preceding the PAM sequence at position 20 has been shown to effect the most significant differences between high- and low-efficiency sgRNA [12]. At this position, guanine is favored and cytosine is disfavored [11,12]. Studies have also shown preference at the preceding nucleotide, base 19, for guanine as well [11]. Additionally, there is a bias for cytosine against guanine at position 16, a preference for adenine in the middle of the sgRNA sequence, and a bias for guanine and against cytosine at position 3 [11,12].

While the nucleotides proximal to the PAM sequence are more important for sgRNA and Cas9 binding specificity than those that are distal to the PAM sequence, it should be noted that the nucleotide directly following the PAM sequence has a strong bias against guanine [12]. Generally, the bases distal to the PAM sequence can tolerate up to as many as 10 mismatches without affecting Cas9 binding [13]. Another position that has been found to mediate efficacy of sgRNA is the variable nucleotide in the *S. pyogenes* PAM sequence, "N"GG. At this position, cytosine and guanine are favored while thymine is disfavored [12,13]. This trend follows the observed unfavorability of T nucleotides occupying the 3′ end of the targeting sequence [11,12,14].

Additionally, the presence of certain sequence motifs have been found to reduce sgRNA efficiency, and, in some cases, block CRISPR/Cas9-mediated gene editing [15]. These short-sequence motifs are located at the 3′ end of the sgRNA targeting sequence 4 bases proximal to the PAM. These bases are located within the "seed" region that plays a crucial role in pairing the sgRNA with the target DNA, mediating specificity and efficiency. The presence of these motifs have been associated with a 10-fold reduction in gene knockout frequencies [15].

The GCC motif involves the presence of a GCC core, a sequence containing DGCC or GCCT in the seeding region. sgRNAs with the GCC motif are potentially inactivated in vivo or unable to recognize and bind the target DNA region, possibly due to the GC-rich sequence interfering with PAM recognition [15]. The TT motif consists of a TT-dinucleotide and one pyrimidine or a sequence of four pyrimidines with at least two T bases. More broadly, the presence of T-rich sequences in the "seeding" region are unfavorable, leading to reduced expression and decreased efficiency; T-rich sequences are more commonly found in the least active sgRNAs compared to the most active sgRNAs [15]. Possible explanations include that RNA polymerase III, which is involved in expression of sgRNAs, terminates at T-rich regions present downstream of the 20-nucleotide variable region [12,15], and T nucleotides interfere with sgRNA loading into Cas9 [10,15]. sgRNAs are transcribed by RNA polymerase III under the control of the U6 promoter in cells. Thus the presence of guanine nucleotides at the 5′ end of the sgRNA is necessary for efficient transcription of the sgRNAs. For sgRNA sequences lacking a guanine at the 5′ end, the addition of 1–2 guanine nucleotides on the 5′ end of an sgRNA lacking the nucleotide can facilitate transcription [3].

Other factors that have been found to mediate sgRNA efficacy include the specific exon and the DNA strand that the sgRNA targets. sgRNAs targeting the last coding exon are less effective and show diminished activity than those targeting earlier exons [10]. This finding could be due to that mutations in the last exon are less disruptive to gene function than those in earlier exons. Additionally, the DNA strand that the sgRNAs targets can impact its efficacy; those targeting the nontranscribed strand are more effective than those targeting the transcribed strand [10].

Another factor that mediates sgRNA and DNA interactions, as well as Cas9 locating the PAM, is the chromatin structure of the DNA [12,13]. Off-target binding of sgRNA and Cas9 occurs primarily in open chromatin regions [13]. This finding is in line with other studies that have shown that CRISPR/Cas9-mediated gene editing is more efficient in euchromatin than in heterochromatin [16], likely because it is easier for the large Cas9 protein to access euchromatic regions. Euchromatic regions of DNA that are not bound by nucleosomes are more accessible to sgRNA–Cas9 complexes [17]. Studies have shown that nucleosomes impede Cas9 binding and cleavage by masking approximately 147 nucleotides, reducing the availability of DNA to Cas9 [17]. Thus CRISPR/Cas9 gene editing does not efficiently target these nucleosome-bound regions in the genome. Determining whether target sites are bound by nucleosomes can be key in maximizing on-target sgRNA activity and minimizing off-target effects. The DNA regions typically masked by nucleosomes are transiently exposed during replication, remodeling, and transcription. Nevertheless, these transient nucleosome displacement events are not sufficient for restoring Cas9 activity in vivo [17]. However, the use of chromatin remodeling enzymes, such as yChd1, have been shown to allow Cas9 to access regions of chromatin typically masked by nucleosomes [17].

sgRNA secondary structures impede formation of active Cas9–sgRNA complexes

Low sgRNA activity can result from defects in forming functional Cas9–sgRNA complexes or from the failure of sgRNA in recognizing the target sites [18]. A functional Cas9–sgRNA complex requires proper interactions between sgRNA and Cas9 as well as sgRNA and the target DNA. One factor that impedes these interactions, and thus Cas9-mediated cleavage, is the formation of secondary structures, such as hairpins, within the sgRNA. These hairpins can result from internal interactions within the crRNA region or between the crRNA and other parts of the sgRNA. Particularly, the backbone region in the sgRNA is more susceptible to form hairpin structures with the variable region [18]. These secondary structures form concurrently during transcription [18]. Although the formation of secondary structures does not prevent binding of the sgRNA to Cas9, it interferes with sgRNA interaction with the target DNA. Thus they form Cas9–sgRNA complexes with low endonuclease activity and deplete the amount of Cas9 available to interact with active sgRNA, ultimately diminishing the potency [18]. sgRNAs with weak secondary structures are more efficient than those with strong secondary structures, likely because sgRNA has more conformational freedom [16]. Particularly, cleavage efficiencies are highest if Gibbs free energy of the sgRNA is between −2 and 0 kJ/mol [16]. Although significant secondary structure formation can impede Cas9 activity, weak secondary structure formation may be necessary to prevent sgRNA interactions with cellular factors such as CTCF, a protein involved in genome organization [18] that can obstruct Cas9 activity [16]. One method of overcoming low sgRNA activity due to formation of secondary structures is disruption of these hairpin structures by heating and then cooling these sgRNAs. This method can lead to increases in CRISPR/Cas9 activity up to 7-fold [18]. Online tools can be used to predict the susceptibility of different sgRNA sequences in forming these secondary structures that interfere with activity.

sgRNA and Cas9 concentration influences off-target activity

Specificity of CRISPR/Cas9 activity is dictated by reaction conditions including the amount of sgRNA and Cas9 present. High enzyme concentrations can lead to increased off-target activity [8]. However, sufficient Cas9 and sgRNA concentrations are needed to mediate efficient gene knockout. The optimal concentrations are also dependent on the potency of the sgRNA [19], which is dependent on characteristics of both the sgRNA and DNA. More potent sgRNAs are able to achieve gene knockout at lower concentrations of Cas9 and sgRNA [19]. Studies have also shown that knockout efficiency in cells is primarily determined by the potency of sgRNA and not on the copy number of the target gene [19]. This observation is likely due to the fact that the amount of Cas9/sgRNA complexes likely exceeds the copy number of the gene.

Strategies to minimize sgRNA off-target activity

One method of minimizing off-target activity involves changing the way DSBs are induced in the region of interest. In the wildtype system, Cas9 induces a DSB at the target sequence mediated by the sgRNA and is repaired through the NHEJ or HDR pathway. In the absence of a repair template, the default NHEJ repair mechanism predominates and frequently leads to indel mutations. In contrast, single-stranded breaks in DNA are repaired in the absence of a template and without any significant indel mutations through the high-fidelity base-excision repair pathway [1,20].

Rather than using the wildtype Cas9 to generate a DSB, an alternative method of generating DSBs necessary for genomic engineering using the CRISPR/Cas9 system involves a double-nicking method. The double-nicking method does not introduce breaks in the target sequence simultaneously, but rather, single-stranded nicks on each strand of DNA Fig. 4.3. This method involves using two sgRNA sequences with a mutant D10A Cas9 nickase (Cas9n). Unlike the wildtype Cas9, the Cas9n generates a single-stranded break. The two sgRNA sequences, one for the sense and one for the antisense strand of DNA, mediate the specificity of the single-stranded breaks. As two sgRNA sequences need to be correctly base paired at the target site to generate a DSB, twice the number of complementary base pairing needs to occur, leading to an increase in specificity and a decrease in off-target mutagenesis. This approach has been shown to reduce off-target activity by 50- to 1000-fold in cell lines, while maintaining high on-target efficiencies comparable to wildtype Cas9 [20]. This strategy, however, can still exhibit off-target

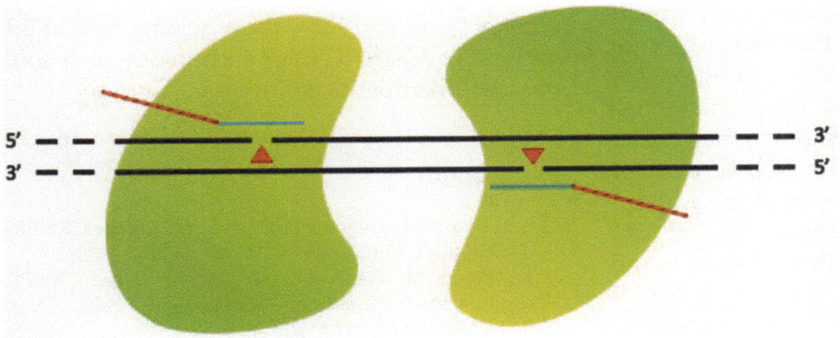

FIGURE 4.3 Use of mutant D10A Cas9 nickases (*green*) induce two single-stranded nicks (*red arrows*) on each DNA strand, each with a unique sgRNA sequence. Generation of two single-stranded nicks rather than a double-stranded break with wildtype Cas9 reduces off-target activity.

activity if both nickases are able to exhibit activity at other DNA regions [20]. Additionally, the Cas9n may still retain residual double-nicking abilities, leading to some background off-target activity [20]. Another limitation in using the paired nicking strategy is that two sgRNAs are needed in the system, and consequently, two PAM sequences, one on the sense and antisense DNA strands, must be present at the target site. This requirement may limit the number of target sites able to undergo CRISPR/Cas9-mediated gene editing.

The double-nicking method can be combined with use of a truncated sgRNA sequence to further improve specificity and efficiency. While extending the length of the sgRNA variable sequence to 20 or 21 nucleotides increases the amount of complementary base paring with the target DNA, the specificity of the Cas9 does not improve. These sgRNAs are actually less efficient, possibly due to interference with Cas9 activity or due to the need for additional RNA processing [7,11,20]. Conversely, truncated sgRNAs with 17 or 18 nucleotide tracrRNA sequences are able to be as stable and efficient as traditional sgRNAs, while suppressing the amount of off-target activity by 5000-fold [21]. The shortest length of the variable sequence for effective sgRNA function has been shown to be 17 nucleotides [22]. These sgRNAs are truncated at the 5′ end of the variable sequence. Truncated sgRNA are more sensitive to base pair mismatches than the full-length sequences, generally tolerating only 1–2 nucleotides mismatches between the RNA and DNA [21]. The reduced tolerance for mismatches is likely due to the binding energy requirements for sgRNA and DNA interaction [21]. An important consideration when using truncated sgRNAs is that its knockout potency differs between cell lines in vivo due to different endogenous levels of Cas9 expression [22]. The 17- and 20-nucleotide sgRNAs were found to have identical knockout potency in 293 T cells with higher levels of Cas9 expression; however, the truncated sgRNAs were less efficient in iPSCs/iMSCs, which have lower levels of Cas9 expression [22].

Another method to improve CRISPR/Cas9 efficiency is by simultaneously using different sgRNAs that all target the same site. Since sgRNAs can exhibit various levels of potency and efficiency, many potential sgRNAs can be designed that all target a specific site. Using sgRNAs with overlaps of several nucleotides has been shown to exhibit improved ssODN-mediated HDR efficiency with no decrease in the frequency of DSBs in gene knock-ins [23].

Conclusion

Careful design of sgRNAs is crucial for optimizing CRISPR/Cas9 activity. Minimizing off-target activity is another critical step to ensure the effective use of this powerful gene-editing tool. Characteristics of the sgRNA including its homology to different sites on the genome, the nucleotides at different locations within the sgRNA, its propensity to form secondary structures, and characteristics of the target DNA site all affect efficacy and potency of sgRNAs. Strategies to reduce off-target activity have included the use of a paired nickase strategy that requires two sgRNAs to mediate increased specificity for a target site and the use of truncated sgRNAs that are less tolerant of base pair mismatches at off-target sites. Further elucidation of sgRNA characteristics to minimize off-target activity and improve efficiency is needed to enhance CRISPR/Cas9 technology as a tool for novel therapeutic strategies.

References

[1] F.A. Ran, et al., Genome engineering using the CRISPR-Cas9 system, Nat. Protoc. 8 (11) (2013) 2281–2308.
[2] X. Wu, et al., Genome-wide binding of the CRISPR endonuclease Cas9 in mammalian cells, Nat. Biotechnol. 32 (7) (2014) 670–676.

[3] H. Kim, J.S. Kim, A guide to genome engineering with programmable nucleases, Nat. Rev. Genet. 15 (5) (2014) 321–334.

[4] M.L. Hefferin, A.E. Tomkinson, Mechanism of DNA double-strand break repair by non-homologous end joining, DNA Repair. (Amst.) 4 (6) (2005) 639–648.

[5] N. Saleh-Gohari, T. Helleday, Conservative homologous recombination preferentially repairs DNA double-strand breaks in the S phase of the cell cycle in human cells, Nucleic Acids Res. 32 (12) (2004) 3683–3688.

[6] M. Jinek, et al., Structures of Cas9 endonucleases reveal RNA-mediated conformational activation, Science 343 (6176) (2014) 1247997.

[7] X. Ren, et al., Enhanced specificity and efficiency of the CRISPR/Cas9 system with optimized sgRNA parameters in Drosophila, Cell Rep. 9 (3) (2014) 1151–1162.

[8] P.D. Hsu, et al., DNA targeting specificity of RNA-guided Cas9 nucleases, Nat. Biotechnol. 31 (9) (2013) 827–832.

[9] Y. Lin, et al., CRISPR/Cas9 systems have off-target activity with insertions or deletions between target DNA and guide RNA sequences, Nucleic Acids Res. 42 (11) (2014) 7473–7485.

[10] T. Wang, et al., Genetic screens in human cells using the CRISPR-Cas9 system, Science 343 (6166) (2014) 80–84.

[11] M.A. Moreno-Mateos, et al., CRISPRscan: designing highly efficient sgRNAs for CRISPR-Cas9 targeting in vivo, Nat. Methods 12 (10) (2015) 982–988.

[12] J.G. Doench, et al., Rational design of highly active sgRNAs for CRISPR-Cas9-mediated gene inactivation, Nat. Biotechnol. 32 (12) (2014) 1262–1267.

[13] C. Kuscu, et al., Genome-wide analysis reveals characteristics of off-target sites bound by the Cas9 endonuclease, Nat. Biotechnol. 32 (7) (2014) 677–683.

[14] H. Xu, et al., Sequence determinants of improved CRISPR sgRNA design, Genome Res. 25 (8) (2015) 1147–1157.

[15] R. Graf, et al., sgRNA sequence motifs blocking efficient CRISPR/Cas9-mediated gene editing, Cell Rep. 26 (5) (2019) 1098–1103. e3.

[16] K.T. Jensen, et al., Chromatin accessibility and guide sequence secondary structure affect CRISPR-Cas9 gene editing efficiency, FEBS Lett. 591 (13) (2017) 1892–1901.

[17] M.A. Horlbeck, et al., Nucleosomes impede Cas9 access to DNA in vivo and in vitro, Elife 5 (2016).

[18] S.B. Thyme, et al., Internal guide RNA interactions interfere with Cas9-mediated cleavage, Nat. Commun. 7 (2016) 11750.

[19] G. Yuen, et al., CRISPR/Cas9-mediated gene knockout is insensitive to target copy number but is dependent on guide RNA potency and Cas9/sgRNA threshold expression level, Nucleic Acids Res. 45 (20) (2017) 12039–12053.

[20] F.A. Ran, et al., Double nicking by RNA-guided CRISPR Cas9 for enhanced genome editing specificity, Cell 154 (6) (2013) 1380–1389.

[21] Y. Fu, et al., Improving CRISPR-Cas nuclease specificity using truncated guide RNAs, Nat. Biotechnol. 32 (3) (2014) 279–284.

[22] J.P. Zhang, et al., Different effects of sgRNA length on CRISPR-mediated gene knockout efficiency, Sci. Rep. 6 (2016) 28566.

[23] D.E. Jang, et al., Multiple sgRNAs with overlapping sequences enhance CRISPR/Cas9-mediated knock-in efficiency, Exp. Mol. Med. 50 (4) (2018) 16.

CHAPTER

5

CRISPR-mediated dense mutagenesis: a tool for rational targeting of multiprotein complexes and the noncoding genome

Falak Sher

Center for Translational & Computational Neuroimmunology, Taub Institute for Research on Alzheimer's Disease and the Aging Brain, Department of Neurology, Columbia University Medical Center, Columbia University, New York, NY, United States

Introduction

Multiprotein complex formation creates the backbone of major cellular tasks, including transcription, translation, organelle and molecule trafficking, energy production, protein folding, cytokinesis, and signaling. Given their versatile application throughout the cell, these multiprotein complexes are promising drug targets [1–4]. The key interfaces of subunits within a complex can potentially be disturbed or destabilized by a small molecule agent that can penetrate the cell [5–9]. Hence, the fine-tuning of disease-relevant complex using small molecules may provide a large number of opportunities to unlock the potential of protein–protein interactions (PPIs) as drug targets [10]. However, mapping the key interfaces of subunits of target complex in relation to its role in disease and cellular fitness can be challenging [11] due to the lack of any robust biophysical techniques capable of measuring PPIs.

Clustered regularly interspaced short palindromic repeats (CRISPR) tiling, a more recent application of CRISPR-based genome editing, offers a comprehensive dense mutagenesis that allows for the systematic investigation of the critical interfaces important for PPI. Three-dimensional structure information of the identified interfaces can inform the rational design of therapeutic small molecules to be utilized for the fine-tuning of disease-associated protein complex. The newly reported method can also uncouple the desired phenotype from other complex functions [12].

CRISPR tiling has also been used to interrogate genomic regions that do not code for proteins including promoters, enhancers, and repressors [13–17]. This approach relies on a set of overlapping, tiled single guide RNA

CRISPR Genome Surgery in Stem Cells and Disease Tissues
DOI: https://doi.org/10.1016/B978-0-12-817876-8.00004-8

(sgRNA) sequences, that mutate virtually every nucleotide in order to interrogate their significance in terms of function (e.g., gene repression and cell fitness) in a pooled manner [18]. CRISPR tiling is highly amenable to any protein complex or noncoding genomic regions that may potentially be manipulated as a target for future drug therapy.

In this chapter, we first focus on the principle and the recent advances in CRISPR tiling. We then discuss key aspects of this approach to identify the critical interfaces within protein complex. We also discuss the available tools and specific considerations for the data analysis and interpretation.

Rational targeting of multiprotein complexes using CRISPR-mediated dense mutagenesis

The double-strand DNA breaks created by CRISPR-Cas9 in the absence of homology donors are repaired by classical error-prone nonhomologous end joining and microhomology-mediated end-joining DNA repair pathways, which generally results in either frameshift or in-frame mutations (indels). Frameshift indels result in the loss of protein function, while in-frame indels result in the gain or loss of a short stretch of amino acid residues (AAs); the latter may or may not impact function depending on where the indel occurs in the protein [19]. For example, proteins with small in-frame indels in noncritical regions will likely retain their function. Conversely, the in-frame indel in essential residues of a protein will result in a complete loss-of-function as a result of this disruption in an important structural component of the protein (Fig. 5.1). Therefore applying large sgRNA tiling libraries (CRIPSR-tiling) can comprehensively disrupt the coding sequences to distinguish essential AAs that directly contribute to cellular phenotypes including cell fitness. Accordingly, a single pooled screen experiment using thousands of sgRNAs, tiling a group of genes coding a

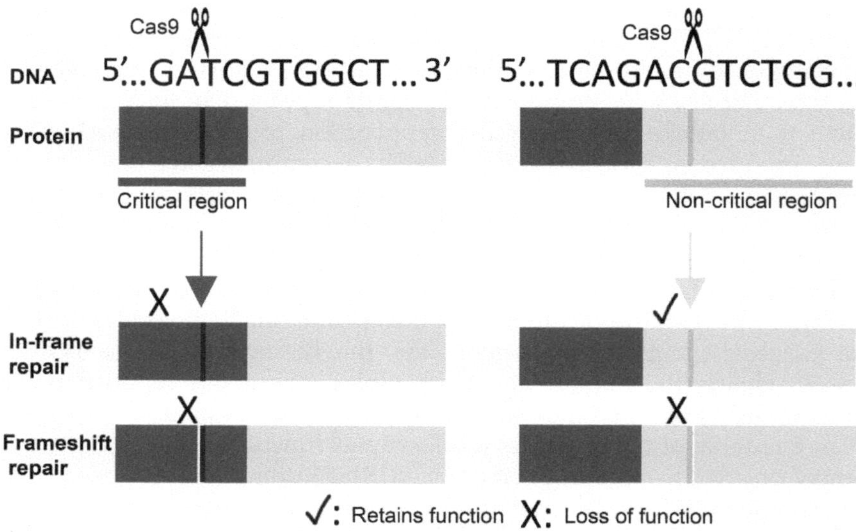

FIGURE 5.1 In-frame indel model. In-frame indel in critical region (AA) of a protein will result into loss-of-function of a protein (left panel), whereas in-frame indel in noncritical region is most likely to be tolerated (right panel).

protein complex, can systematically disrupt the function of virtually all the AAs within a multiprotein complex, to identify the AA essential for PPI attributed to a phenotype of interest and cytotoxicity. Until now this approach has been successfully used to target single or multiple genes in a high-throughput manner to identify pharmaceutically important protein domains [12,20–22].

Recent research has combined CRISPR-tiling with positive selection to identify small molecule drug sites within a protein [23] and to predict clinical drug resistance with improved accuracy [24]. More recently, the methodology led the identification of key protein interfaces within nucleosome remodeling and deacetylase (NuRD) chromatin complex where in-frame alleles resulted in loss-of-function due to destabilization or altered function of the subunit [12]. Interestingly, the study identified three AAs within CHD4, a subunit of NuRD, where an in-frame deletion uncoupled its requirement for cell fitness from gene repression, both in vitro as well as in vivo. Given this information, the identified three AAs were regarded an ideal pharmacologic target for gene reactivation without causing significant cellular toxicity. Below we highlight the key aspects of CRISPR-tiling to identify the functional protein coding sequences within a multiprotein complex.

CRISPR-tiling screen workflow

In order to identify the key interfaces within a protein complex, a typical CRISPR-tiling screen includes(1) following workflow; (2) the design, synthesis, and construction of lentiviral library, saturating sgRNAs at coding DNA sequences of target protein complex; (3) viral transduction into target cells; (4) selection of a phenotype relevant to the disease or biological process in question; (5) readout of sgRNAs by next-generation sequencing (NGS); and (6) analysis of NGS data (Fig. 5.2).

CRISPR-tiling library design, synthesis, and construction of lentivirus

When designing a CRISPR-tiling library, there are many key questions to consider, for example, which cell line to use, how many positive and negative control sgRNAs should be included, what should be the threshold to filter out low specificity sgRNAs from the pool, and how to confirm the editing. A typical CRISPR-tiling library requires identification of coding sequences of the target complex and a choice of CRISPR nuclease (Cas9), for example, commonly used *Streptococcus pyogenes* Cas9 with protospacer adjacent motif (PAM)-NGG or its variants, such as Cas-NG and xCas9 [25,26] which can recognize a broad range of PAM sequences including -NG, -GAA, and -GAT [27]. These newly developed *S. pyogenes* Cas9 variants can increase the density of sgRNAs over target regions as they are not restricted by -NGG PAM availability in the target sequence.

The candidate sgRNAs pool includes all the target sequences followed by a PAM sequence, for example, in the case of *S. pyogenes* Cas9 every 20-mer sequence upstream of -NGG PAM on the sense and antisense strands of coding regions of target genes. A number of cloud or command line computational tools (e.g., ProTiler [22], sgTiler [28], CRISPOR [29], CRISPR library designer [30]) are available to identify candidate sgRNAs. These tools can distinguish all available sgRNAs, for example, those which map uniquely, across the entire coding region of interest. Additionally, they can rank sgRNAs according to different scores [30–33] that evaluate potential off-targets in the genome of interest and predict the effect of on-target activity. Prediction of secondary structures in the candidate sgRNAs can

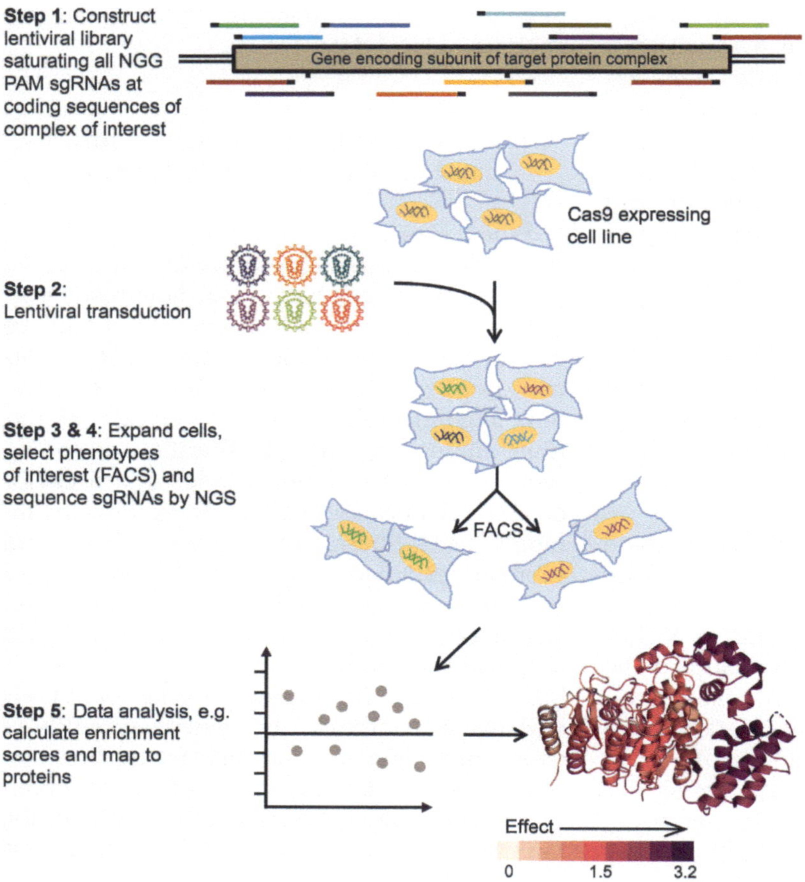

FIGURE 5.2 Schematic representation of CRISPR-tiling screen.

improve their cleavage efficiency [34]. Comparative studies [35] suggest that specificity scores of sgRNAs could be assay dependent. For example, a Doench score [32] is recommended when sgRNAs are expressed inside cells from exogeneous promoters such as the U6 promoter, whereas the Moreno-Mateos scores [33] should be taken into account when guide RNAs are transcribed in vitro from the T7 RNA polymerase promoter [29]. Moreover, understanding genetic variations in the target cell line can further improve the prediction of off-target effects within the candidate sgRNA pool [36]. It is important to distribute the sgRNAs in a way that they are not sporadically clustered, that is, they span across the whole target region instead of leaving gaps in between. The library should also include abundant negative control sgRNAs (sgRNAs expected to have no biological impact) as well as positive control sgRNAs (e.g., top ranked sgRNAs validated in a separate arrayed screen). Previously published CRISPR-tiling libraries [12] suggest that including 4.5% of negative control sgRNAs should be enough to confer the effect of specific sgRNAs on biological phenotypes.

After designing the library, the next step is to synthesize and clone the library into the lentiviral vector. Oligonucleotides can be synthesized using array synthesis [13,37,38] and cloned as a pool into the lentiviral vector of choice. Sufficient colonies should be isolated to ensure ~90x library coverage. After cloning, the plasmid library should be deep sequenced to ensure the representation of each sgRNA in the pool. Specifically, HEK293T cells are the best choice to produce supernatant containing synthesized lentivirus.

Lentiviral transduction and cell expansion

Many continuous laboratory cell lines have an aberrant number of chromosomes and focal gene amplification due to genomic instability [39]. Cell lines with more than two copies of target genes can negatively affect knockout efficiency of Cas9. Therefore karyotype of the cell line of interest should be determined to infer the gene copy number before starting the screen. It may be necessary to derive a single cell clone from a pool of Cas9-expressing cells with high expression of Cas9 and two confirmed copies of the target genes. The Cas9 cleavage efficiency can be confirmed using Cas9 activity assay [31] in target cell line, prior to the start of screen.

After low multiplicity of infection transduction (0.3), cells should undergo expansion for ~2 weeks before beginning phenotype-based selection, so that there is sufficient time for gene editing and downstream changes in transcription and protein expression. During this expansion, frameshift indels should tend to abolish protein function while alleles with in-frame indels may or may not impact function depending on the location of the indel [12,19,20,40].

Phenotype selection and next-generation sequencing

The cells of a particular phenotype, resulting from an enrichment of genetic perturbations, can be isolated using flow cytometry, and then sorted based on changes in a particular phenotype (e.g., heightened phenotype vs lowered phenotype). Genomic DNA should be collected from (1) unsorted, (2) heightened, and (3) lowered phenotype populations. The viral integrant serves as a tag for readout of which sgRNA construct was delivered to a particular cell. Using genomic DNA, NGS libraries are then prepared for deep sequencing by a two-step polymerase chain reaction (PCR) [41]. First, the PCR amplifies the genome-integrated CRISPR constructs. This step is critical, and requires an adequate amount of genomic DNA to represent the desired number of cells. Next, the PCR introduces the sequencing barcodes and unique molecular identifiers [42–44] to reduce the artifacts that may skew quantification of amplicon sequencing.

Analysis of CRISPR-tiling screen

The key steps in the analysis of CRISPR-tiling screen include (1) counting of sgRNAs in each sequenced population (FASTQ file) along with the calculation of functional scores of individual sgRNAs; (2) mapping of functional scores associated with each sgRNA to genome, transcript, and protein; and (3) displaying the data on structure of protein if available, or, alternatively, the identified AA sequence can be nominated for the purification of crystal structure. Functional scores of sgRNAs can be calculated as logtwofold change of guide representation in the sequenced populations. For example, sgRNAs dropout from library to presort represents toxicity, that is, perturbation of target AAs resulted in cell death. Computational pipelines for example, CRISPRO [45], ProTiler [22], and PASTMUS [46] are available for the robust analysis of CRISPR-tiling screens. CRISPRO can analyze and visualize the CRISPR-tiling data sets. The pipeline can precisely count sgRNA abundance in various samples, can map functional scores associated with sgRNA to genome, transcripts, and protein coordinates and structures. Similarly, ProTiler and PASTMUS pipelines can precisely pinpoint critical AAs of proteins from CRISPR-tiling screen NGS data.

Future directions and areas for improvement

CRISPR-tiling targets all the sequences covered by a PAM motif in the region of interest. However, one possible pitfall encountered with all sgRNA approaches is the potential for the inclusion of inactive sgRNAs and sgRNAs with significant off-target effects, which can confound the screen results. Moreover, guides targeting repetitive DNA regions have been associated with false positives [20,47,48], and as such, it is recommended to filter out inactive and/or low specificity guides, as well as guides targeting repetitive DNA regions when designing CRISPR-tiling libraries. This technique, however, then also compromises the overall resolution of the library. To improve the reliability of the data obtained from CRISPR-tiling screens, first an improvement in the library design tools—in terms of specificity and activity of the sgRNAs as well as the ability to predict genome editing outcome—must take place. Several limitations to the CRISPR-tiling technique remain as well. The resolution of the technique's screens is restricted to the availability of PAM sequences (e.g., NGG for SpCas9). This problem may be partially addressed via the use of newly developed orthologs and variants of Cas nucleases with alternative PAM options [25,26], but has still yet to be ameliorated completely. Another potential confounding factor in the CRISPR-tiling library design includes disease and trait-associated genetic coding variants identified in genome-wide association studies. By incorporating genetic variants in the design of the library, we are once again able to improve the outcome of CRISPR-tiling screens. Newly developed CRISPR library design tools [47] can incorporate alternative sgRNAs based on haplotype structure, whole genome sequencing, or custom list of variants. Such tools can facilitate the design of variant-aware CRISPR-tiling libraries, which can broaden the quality of the screen data.

Conclusions

Pooled CRISPR-tiling is a powerful tool to pinpoint the most critical AAs within a protein complex responsible for a particular phenotype, for example, drug sensitivity and cell fitness. Structure determination of identified AAs can lead to rational design of small drug-like molecules for therapeutic modulation of PPIs. Advancement in CRISPR technologies will continue to improve resolution, efficiency, and specificity of CRISPR-tiling libraries. As the field of genome editing continues to expand and develop, so too we anticipate that CRISPR-tiling strategies will evolve rapidly to extensively map disease-associated multiprotein complexes.

References

[1] D.P. Ryan, J.M. Matthews, Protein-protein interactions in human disease, Curr. Opin. Struct. Biol. 15 (4) (2005) 441–446.

[2] J.A. Gerrard, C.A. Hutton, M.A. Perugini, Inhibiting protein-protein interactions as an emerging paradigm for drug discovery, Mini Rev. Med. Chem. 7 (2) (2007) 151–157.

[3] P. Chene, Drugs targeting protein-protein interactions, ChemMedChem 1 (4) (2006) 400–411.

[4] I.S. Moreira, P.A. Fernandes, M.J. Ramos, Hot spots—a review of the protein-protein interface determinant amino-acid residues, Proteins 68 (4) (2007) 803–812.

[5] N.S. Robertson, D.R. Spring, Using peptidomimetics and constrained peptides as valuable tools for inhibiting protein-protein interactions, Molecules 23 (4) (2018) 959.

[6] G. Zhang, J. Andersen, G. Gerona-Navarro, Peptidomimetics targeting protein-protein interactions for therapeutic development, Protein Pept. Lett. 25 (12) (2018) 1076–1089.

[7] L. Du, et al., Peptidomimetics for targeting protein-protein interactions between DOT1L and MLL Oncofusion proteins AF9 and ENL, ACS Med. Chem. Lett. 9 (9) (2018) 895–900.

[8] L. Mabonga, A.P. Kappo, Protein-protein interaction modulators: advances, successes and remaining challenges, Biophys. Rev. 11 (4) (2019) 559–581.

[9] O. Slater, B. Miller, M. Kontoyianni, Decoding protein-protein interactions: an overview, Curr. Top. Med. Chem. 20 (10) (2020) 855–882.

[10] L.M. Stevers, et al., Modulators of 14-3-3 protein-protein interactions, J. Med. Chem. 61 (9) (2018) 3755–3778.

[11] J.A. Wells, C.L. McClendon, Reaching for high-hanging fruit in drug discovery at protein-protein interfaces, Nature 450 (7172) (2007) 1001–1009.

[12] F. Sher, et al., Rational targeting of a NuRD subcomplex guided by comprehensive in situ mutagenesis, Nat. Genet. 51 (7) (2019) 1149–1159.

[13] M.C. Canver, et al., BCL11A enhancer dissection by Cas9-mediated in situ saturating mutagenesis, Nature 527 (7577) (2015) 192–197.

[14] C.P. Fulco, et al., Systematic mapping of functional enhancer-promoter connections with CRISPR interference, Science 354 (6313) (2016) 769–773.

[15] T.S. Klann, et al., CRISPR-Cas9 epigenome editing enables high-throughput screening for functional regulatory elements in the human genome, Nat. Biotechnol. 35 (6) (2017) 561–568.

[16] N.E. Sanjana, et al., High-resolution interrogation of functional elements in the noncoding genome, Science 353 (6307) (2016) 1545–1549.

[17] D.R. Simeonov, et al., Discovery of stimulation-responsive immune enhancers with CRISPR activation, Nature 549 (7670) (2017) 111–115.

[18] M.C. Canver, D.E. Bauer, S.H. Orkin, Functional interrogation of non-coding DNA through CRISPR genome editing, Methods 121–122 (2017) 118–129.

[19] J. Shi, et al., Discovery of cancer drug targets by CRISPR-Cas9 screening of protein domains, Nat. Biotechnol. 33 (6) (2015) 661–667.

[20] D.M. Munoz, et al., CRISPR screens provide a comprehensive assessment of cancer vulnerabilities but generate false-positive hits for highly amplified genomic regions, Cancer Discov. 6 (8) (2016) 900–913.

[21] M.A. Erb, et al., Transcription control by the ENL YEATS domain in acute leukaemia, Nature 543 (7644) (2017) 270–274.

[22] W. He, et al., De novo identification of essential protein domains from CRISPR-Cas9 tiling-sgRNA knockout screens, Nat. Commun. 10 (1) (2019) 4541.

[23] J.E. Neggers, et al., Target identification of small molecules using large-scale CRISPR-Cas mutagenesis scanning of essential genes, Nat. Commun. 9 (1) (2018) 502.

[24] L. Ma, et al., CRISPR-Cas9-mediated saturated mutagenesis screen predicts clinical drug resistance with improved accuracy, Proc. Natl Acad. Sci. U.S.A. 114 (44) (2017) 11751–11756.

[25] J.H. Hu, et al., Evolved Cas9 variants with broad PAM compatibility and high DNA specificity, Nature 556 (7699) (2018) 57–63.

[26] H. Nishimasu, et al., Engineered CRISPR-Cas9 nuclease with expanded targeting space, Science 361 (6408) (2018) 1259–1262.

[27] P.D. Hsu, et al., DNA targeting specificity of RNA-guided Cas9 nucleases, Nat. Biotechnol. 31 (9) (2013) 827–832.

[28] M. Ahmed, H.H. Housheng, SgTiler: a fast method to design tiling sgRNAs for CRISPR/Cas9 mediated screening. bioRxiv (2017). https://doi.org/10.1101/217166

[29] J.P. Concordet, M. Haeussler, CRISPOR: intuitive guide selection for CRISPR/Cas9 genome editing experiments and screens, Nucleic Acids Res. 46 (W1) (2018) W242–W245.

[30] F. Heigwer, et al., CRISPR library designer (CLD): software for multispecies design of single guide RNA libraries, Genome Biol. 17 (2016) 55.

[31] J.G. Doench, et al., Rational design of highly active sgRNAs for CRISPR-Cas9-mediated gene inactivation, Nat. Biotechnol. 32 (12) (2014) 1262–1267.

[32] J.G. Doench, et al., Optimized sgRNA design to maximize activity and minimize off-target effects of CRISPR-Cas9, Nat. Biotechnol. 34 (2) (2016) 184–191.

[33] M.A. Moreno-Mateos, et al., CRISPRscan: designing highly efficient sgRNAs for CRISPR-Cas9 targeting in vivo, Nat. Methods 12 (10) (2015) 982–988.

[34] T. Wang, et al., Genetic screens in human cells using the CRISPR-Cas9 system, Science 343 (6166) (2014) 80–84.

[35] M. Haeussler, et al., Evaluation of off-target and on-target scoring algorithms and integration into the guide RNA selection tool CRISPOR, Genome Biol. 17 (1) (2016) 148.

[36] B.P. Kleinstiver, et al., High-fidelity CRISPR-Cas9 nucleases with no detectable genome-wide off-target effects, Nature 529 (7587) (2016) 490–495.

[37] N.E. Sanjana, O. Shalem, F. Zhang, Improved vectors and genome-wide libraries for CRISPR screening, Nat. Methods 11 (8) (2014) 783–784.

[38] O. Shalem, et al., Genome-scale CRISPR-Cas9 knockout screening in human cells, Science 343 (6166) (2014) 84–87.

[39] M.R. Stratton, P.J. Campbell, P.A. Futreal, The cancer genome, Nature 458 (7239) (2009) 719–724.

[40] K.F. Donovan, et al., Creation of novel protein variants with CRISPR/Cas9-mediated mutagenesis: turning a

screening by-product into a discovery tool, PLoS One 12 (1) (2017) e0170445.

[41] L.A. Miles, R.J. Garippa, J.T. Poirier, Design, execution, and analysis of pooled in vitro CRISPR/Cas9 screens, FEBS J. 283 (17) (2016) 3170–3180.

[42] K. Clement, et al., AmpUMI: design and analysis of unique molecular identifiers for deep amplicon sequencing, Bioinformatics 34 (13) (2018) i202–i210.

[43] I. Kinde, et al., Detection and quantification of rare mutations with massively parallel sequencing, Proc. Natl Acad. Sci. U.S.A. 108 (23) (2011) 9530–9535.

[44] S.R. Kennedy, et al., Detecting ultralow-frequency mutations by duplex sequencing, Nat. Protoc. 9 (11) (2014) 2586–2606.

[45] V.A.C. Schoonenberg, et al., CRISPRO: identification of functional protein coding sequences based on genome editing dense mutagenesis, Genome Biol. 19 (1) (2018) 169.

[46] X. Zhang, et al., PASTMUS: mapping functional elements at single amino acid resolution in human cells, Genome Biol. 20 (1) (2019) 279.

[47] M.C. Canver, et al., Variant-aware saturating mutagenesis using multiple Cas9 nucleases identifies regulatory elements at trait-associated loci, Nat. Genet. 49 (4) (2017) 625–634.

[48] A.J. Aguirre, et al., Genomic copy number dictates a gene-independent cell response to CRISPR/Cas9 targeting, Cancer Discov. 6 (8) (2016) 914–929.

CHAPTER

6

Targeting Alzheimer's disease and related dementias with CRISPR and human pluripotent stem cell technologies

Andrew A. Sproul[1,2]

[1]Taub Institute for Research on Alzheimer's Disease and the Aging Brain, Columbia University, New York, NY, United States [2]Department of Pathology and Cell Biology, Columbia University Irving Medical Center, New York, NY, United States

The numbers are stark. By 2050, 1 in 85 people are predicted to have Alzheimer's disease (AD). More than 400 clinical trials to treat AD have failed. In 2015, the worldwide costs of dementia were estimated to be US$818 billion, equivalent to 1.09% of the global gross domestic product and the 18th largest economy in the world [1–3]. This cost will increase significantly as more people live longer. Beyond the numbers, many of us are personally affected by dementia, whether in our own or our friends' families. My wife and I each lost our paternal grandfather to unclassified dementias, and unfortunately I have no memories of my grandfather since I was very young. Many of us fear that our parents or grandparents will develop dementia, or that even we, ourselves, will as we get older. When I meet a new senior and share that I am an AD researcher, they often say something like "good, hurry up I'm going to need it" as a joke. However, in most cases they are not really joking. In this chapter, we will explore how clustered regularly interspaced short palindromic repeats (CRISPR) and human-induced pluripotent stem cell (hiPSC) technologies will help the development of new therapeutics capable of successfully targeting dementias as well as the health, economic, and psychological tolls that these devastating diseases reap.

Overview of AD and related dementias and how CRISPR technologies may help

Dementia is broadly defined as the loss of cognitive ability in multiple areas, which significantly interferes with daily life. Classical AD, the leading cause of dementia worldwide, is associated with loss of memory and is definitively

CRISPR Genome Surgery in Stem Cells and Disease Tissues
DOI: https://doi.org/10.1016/B978-0-12-817876-8.00002-4

diagnosed postmortem by the presence of extracellular amyloid plaques primarily composed of the Aβ peptide and intracellular neurofibrillary tangles primarily composed of hyperphosphorylated tau protein in the autopsied brain. In addition to plaques and tangles, a third pathology was described by Alois Alzheimer in his initial AD characterization in 1906, namely the presence of intracellular lipid droplets (primarily composed of cholesterol esters) [4]. AD is traditionally classified into two buckets, early-onset AD (EOAD), also known as familial AD (FAD), and late-onset AD (LOAD) which affects people of the age 65 years and above and constitutes the majority of AD. EOAD, which comprises 2%–5% of AD cases, is associated with autosomal dominant, largely 100% penetrant mutations in amyloid precursor protein (APP). Presenilin-1 (PSEN1), or its homolog Presenilin-2 (PSEN2), although some mutations in sortilin-related receptor 1 (SORL1) have also been identified recently as EOAD genetic risk factors [5,6]. No LOAD genetic risk factor is sufficient to cause AD, although some genetic variants such as the apolipoprotein E4 (APOE4) allele or mutations in triggering receptor expressed on myeloid cells 2 greatly increase disease risk [7,8]. The greatest environmental (and overall) risk factor for LOAD is age, and other strong predispositions include cardiovascular risk factors and diabetes [9–12].

The most prevalent model to explain the pathogenesis of AD is the amyloid cascade hypothesis [13], in which accumulation of toxic Aβ peptides sets into motion downstream pathology, including neuroinflammation, synaptic dysfunction, tau dysregulation, and cell death.

As shown in Fig. 6.1, Aβ is derived from sequential cleavage of APP, first by the β-secretase complex containing the protease BACE1, followed by the γ-secretase complex in which PSEN1/2 make up the catalytic core. This amyloidogenic processing is predominantly thought to occur in early endosomes, where the acidic environment promotes BACE1 cleavage of APP. Importantly, as described in more detail below, other potentially toxic species are also produced via β-cleavage of APP, including the C99/β-CTF fragment (which is a substrate for the γ-secretase complex). Alternatively, APP can be cleaved by α-secretase at the plasma membrane, which produces neutral or potentially beneficial cleavage products. The genetics of EOAD (APP, PSEN1/2) strongly support a central role for amyloidogenic APP processing, and, conversely, a rare mutation in APP (A693T, "Icelandic mutation") that diminishes β-cleavage is protective against LOAD [14]. Many (but not all) PSEN1/2 mutations are associated with greater production of longer oligomerogenic Aβ peptides such as Aβ42, relative to more common Aβ40 species [13]. The relative ratio of Aβ42:Aβ40 has been argued to be more critical than absolute levels of Aβ42 [15].

While AD is estimated to account for 60% of all dementia, other related dementias (referred collectively with AD as ADRDs) include frontotemporal degeneration (FTD), Lewy body dementia (LBD), vascular contributions to cognitive impairment and dementia (VCID), and mixed etiology dementias (MED), with many distinct subclasses [16]. FTD, also referred to as frontotemporal lobar degeneration, includes atrophy of the frontal and temporal lobes, is often associated with early behavioral changes, and is pathologically associated with aggregated tau in the majority of cases, although alternatively, TDP-43 or FUS can be aggregated in a spectrum with amyotrophic lateral sclerosis. LBD is associated with α-synuclein inclusions (Lewy bodies) damaging cerebral structures, similar to what is found in the midbrain of Parkinson's disease patients with which LBD is in a clinical continuum. VCID is a heterogeneous collection of dementias resulting from vascular damage and is not associated with aggregated proteins. MED refers to a mixture of different dementia subclasses in one patient, for instance, AD and vascular dementia. This commonly occurs in older patients with dementia, unlike early-onset forms of ADRDs, such as familial FTD, which have cleaner etiologies.

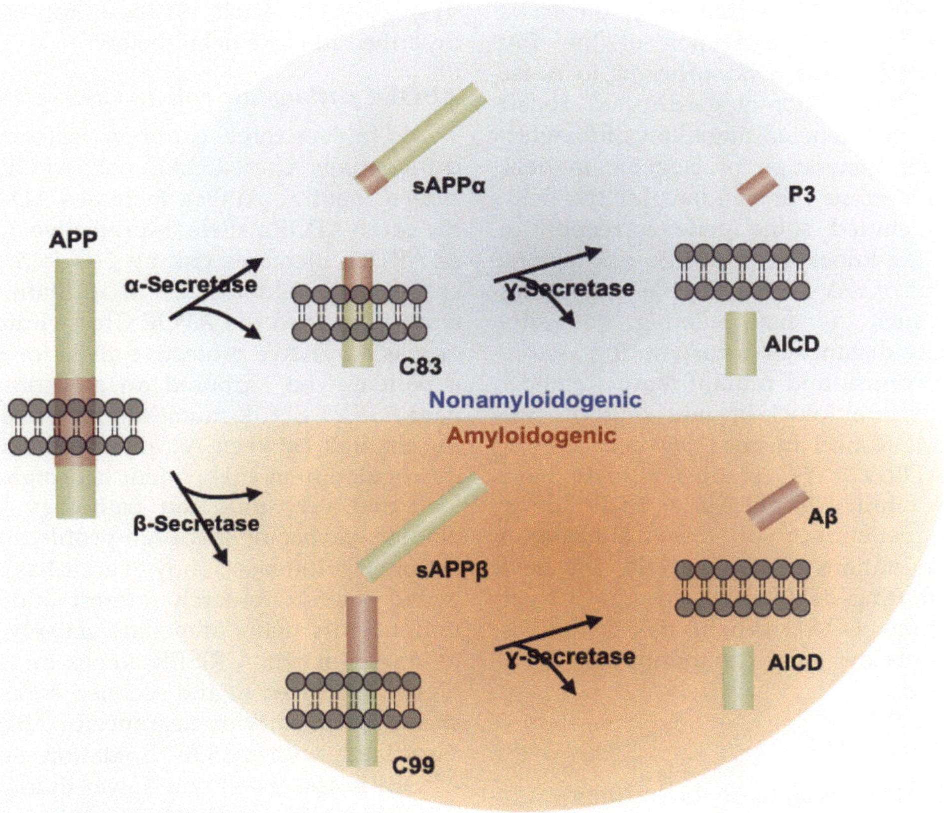

FIGURE 6.1 Model of APP processing. *Source: Courtesy So Yeon Koo.*

Importantly, there are no effective treatments for ADRDs that halt disease progression or reverse pathology, and hundreds of compounds with significant efficacy to cure AD in mouse models have gone on to fail during human clinical trials. Better ADRD disease modeling focusing on human-specific systems such as hiPSC-derived central nervous system (CNS) cell types may provide novel insight into disease pathogenesis to facilitate the development of novel therapeutics [17]. With the advent of HDR-mediated CRISPR genome editing, isogenic systems are now the standard for hiPSC deep disease modeling. Other CRISPR technologies that modulate gene expression—including CRISPR activation (CRISPRa) or CRISPR inhibition (CRISPRi)—enable screening of risk factors, protective genetic factors, or a larger unbiased set of genes, in specific disease-related contexts. Finally, CRISPR technologies also have the potential to be used as a direct therapeutic for ADRDs with prevalidation in hiPSC models or to gene-correct patient hiPSCs prior to personalized regenerative medicine.

CRISPR for disease modeling genetic risk factors in hiPSCs

The first isogenic ADRD hiPSC model was generated a few years before the widespread use of CRISPR, in which zinc finger nucleases were used for gene correction of the Tau A152T mutation, which is sufficient to cause FTD [18]. CRISPR homology-directed repair (HDR)-mediated gene editing allows for significantly easier generation of isogenic models, which have become the standard for the field. This has included some gene correction of mutant alleles, knock-in of genetic risk factors, or knock-out of risk alleles. Simple but elegant strategies such as manipulating cut site-mutation site distance or incorporating a 50:50 mixture of control and mutant template allow for the selective knock-in into one or two alleles via HDR mechanisms [19].

Many ADRD hiPSC studies to date have focused on familial forms AD or FTD, due to the 100% penetrance of these autosomal-dominant mutations. More recently, the field has been shifting its focus to strong, but not fully penetrant, LOAD genetic risk factors, as this represents the most commonly seen form of the disease.

Isogenic hiPSC models of Alzheimer's disease

AD modeling in hiPSCs has begun identifying and strengthening emerging concepts in the AD field. Some of the most important findings, from my perspective, relate to the following: (1) disease pathomechanisms of the APOE4 LOAD risk allele in non-astrocyte CNS populations; (2) discovery of enhanced neurogenesis and maturation in AD hiPSC-derived neural progenitor cells (NPCs), which could have implications for adult neurogenesis or even potentially be indicative of developmental aspects of AD; and (3) the critical role of the APP C99/β-CTF fragment rather than Aβ (or at least exclusively Aβ) as a key driver of some aspects of disease pathogenesis in neurons, including enlarged early endosomes and tau dysregulation, potentially via cholesterol dysregulation. Each of these topics will be described in more detail below.

APOE4 pathogenic role in CNS cell types

APOE has three common isoforms which differentially affect LOAD risk. APOE3 is considered neutral, APOE4 increases AD risk, and the rarer APOE2 allele is protective. One copy of APOE4 increases risk by 2–4-fold, and two copies by 8–12-fold [20]. In addition, a recent study identified the APOE Christchurch mutation as a putative protective allele for EOAD in a patient who harbored an autosomal dominant *PSEN1 E280K* mutation, potentially severing the link between Aβ accumulation and tau dysregulation as this patient had high amyloid load and very mild tau pathology [21]. The obvious caveat of this high-profile study was its low $n = 1$ subjects; however, it has also generated intense research interest and we and undoubtedly many others are actively working to model it via CRISPR knock-in. APOE is highly expressed in and secreted by astrocytes, and is involved with clearance of Aβ, which is associated with APOE lipidation state [22]. APOE4 is associated with lower lipidation, and APOE4/4 hiPSC-derived astrocytes have been shown to have reduced Aβ clearance relative APOE3/3 isogenic control astrocytes [23]. APOE is also the primary cholesterol transporter in the CNS.

Although APOE expression is significantly lower in neurons than astrocytes, APOE has been detected in human CNS neurons [24] and APOE4/4 hiPSC-derived neuronal models from multiple groups demonstrated strong phenotypes relative to isogenic APOE3/3 controls generated by CRISPR gene editing. This includes elevated phospho-tau (pTau), enlarged early endosomes (associated with enhanced amyloidogenic processing of Aβ and observed in AD patients [25]), elevated Aβ42 or both Aβ42 and Aβ40 depending on the study, GABAergic neuron cell death, accelerated neuronal differentiation, and significant gene-expression changes [23,26]. One study argued that these phenotypes were driven by a gain-of-function (GOF) of APOE4, as Aβ, pTau, and GABAergic neuron cell death could be induced in APOE knock-out neurons transfected with an APOE4 but not an APOE3 overexpression construct [26]. There is still significant debate in the AD field if APOE4 primarily acts as GOF or loss-of-function (LOF) relative to APOE3 with regard to increased AD risk.

One key take home point from these neuronal APOE studies is that the cell type which has the highest expression of a gene is not necessarily the cell type that primarily drives pathology from risk variants in that gene. This concept is supported by another recent study looking at SORL1 homozygous knock-out hiPSCs introduced by CRISPR, compared to isogenic controls [27]. Although SORL1 is expressed at significantly higher levels in microglia than cortical neurons [28], enlarged early endosomes were seen in SORL1 knock-out neurons (similar to APOE4/4 neurons), whereas in SORL1 knock-out microglia, early endosomes were similar to those in isogenic controls. This suggests that researchers should not narrowly focus on cell types with the highest expression of a mutant transcript.

Finally, APOE4/4 versus isogenic APOE3/3 hiPSC-derived cells have also been evaluated in several other CNS cell types. APOE4/4 microglia reduced Aβ phagocytosis and increased inflammatory markers relative to APOE3/3 [23]. In an hiPSC blood brain barrier (BBB) model, APOE4/4 pericytes were associated with recruitment of Aβ to the BBB through an NFAT-dependent pathway [29]. Overall, this suggests that APOE4 contributed to AD pathogenesis by affecting disease-relevant processes in multiple cell types in a myriad of ways.

Enhanced neurogenesis and maturation in AD hiPSCs

Multiple studies suggest that adult hippocampal neurogenesis (AHN) in the dentate gyrus of humans, which evidence suggests is maintained throughout life [30], may be negatively impacted in AD. Such evidence includes staining of autopsied brain tissue [31,32] and reduced proliferative capacity and neurogenesis of glial progenitor cells grown from recently deceased patients thought to be due to reduced β-catenin signaling [33], although conflicting studies exist in the literature [34,35]. Many mouse models of AD have shown decreased AHN at the onset of disease pathology [36]. In some models such as the J20 mouse (*APP*$^{Swe/Ind}$ double transgenic), APOE4 knock-in, or APOE knock-out mice, enhanced neurogenesis is seen in young adult mice and then it reduced as the animal's age increases [34,37–39]. This is consistent with an earlier study which indicated that Aβ oligomers enhance neurogenesis when applied to rodent neural stem cells [40]. Overall, this suggests a model where excessive neurogenesis could exhaust the adult neural stem pool. In addition, the APP intracellular domain fragment that is also produced by APP processing (Fig. 6.1) has been shown to inhibit neurogenesis [41,42]. It is thus difficult to predict a priori how neurogenesis might be affected in AD hiPSCs.

The first study to demonstrate enhanced neurogenesis in AD NPCs was a study from my postdoctoral research using retroviral reprogrammed PSEN1 mutant hiPSCs and control hiPSCs from unaffected relatives (FAD1 [A256E] and FAD4 [M146L]) [43,44]. The enhanced early neurogenesis we observed (increased NCAM + cells, patches of early complex neurites in AD NPCs) was not described mechanistically. Recently, however, a mechanistically detailed study was published looking at neurogenesis in APOE4/4 versus isogenic APOE3/3 NPCs generated via CRISPR, as well as sporadic AD hiPSCs versus age-matched controls [45]. AD NPCs from either APOE4/4 or sporadic cases showed accelerated neurogenesis and electrophysiological maturation and a distinct molecular signature, driven at least in part by less activity of the transcriptional repressor restrictive element-1 silencing transcription factor (REST)/neuron-restrictive silencing factor. REST is a major inhibitor of neurogenesis, although its activity has also been proposed to have a protective function in post-mitotic neurons in the context of AD [46,47]. AD NPCs have reduced REST localized in the nucleus as well as disrupted nuclear lamina. Accelerated neurogenesis and maturation phenotypes could be rescued by overexpression of REST and introduced into control NPCs by interfering with REST function. Additional reports using non-isogenic PSEN1 mutant hiPSC models also support enhanced neurogenesis in PSEN1 mutant cells (monolayer or organoid) relative to controls, although reductions in Notch signaling, which is also a target of γ-secretase cleavage, was proposed as the primary mechanism [48,49].

The enhanced neurogenesis seen in hiPSC models by multiple groups is consistent with a model in which early exhaustion of the NPC pool during AHN leads to neurogenesis deficits later during AD disease progression. However, hiPSCs more truly reflect developmental neurogenesis rather than adult neurogenesis. Changes in the neural differentiation and maturation of AD hiPSCs raises the attractive hypothesis that AD risk factors might have developmental effects that could influence disease risk in the aged adult. One example of such a possibility would be a developmental-driven change in the excitatory and inhibitory balance, which is thought to be disrupted in AD [50]. The study demonstrating a loss of GABAergic neurons in APOE4/4 versus APOE3/3 neural cultures is in line with this hypothesis [26]. In addition, changes in neurogenesis kinetics could also influence interpretations of AD hiPSC-derived neuronal models, as phenotypes could result at least in part from AD cells being at different developmental stages. Thus it is critical to fully understand the mechanism(s) of accelerated neuronal differentiation in AD hiPSC-derived NPCs, and isogenic models created via CRISPR will be very helpful in this regard.

The critical role of C99/β-CTF in AD pathology and links with cholesterol dysregulation

Evidence that the C99/β-CTF fragment of APP, also produced by amyloidogenic signaling (Fig. 6.1), is a crucial driver of AD pathology has been accumulating and has been strongly supported by AD hiPSC models. For example, two independent groups have independently demonstrated that APP mutant neurons (APP triple; *APP*^{London} *V717I*, Trisomy 21; non-isogenic) have increased phospho and total tau relative to control neurons, which is blocked by β-secretase but not γ-secretase inhibitors [51,52]. In fact, the latter study demonstrated that γ-secretase inhibitors actually elevate phospho and total tau. Although AD clinical trials using γ-secretase inhibitors were halted due to deleterious effects such as skin cancer, which was attributed to off-target inhibition of Notch [53], the effects of γ-secretase inhibitors on tau suggest this would not be a good therapeutic strategy for AD regardless. Overall, these data support that a β-cleavage product of APP such as C99 rather than Aβ itself is responsible for tau dysregulation, as γ-secretase inhibition would block Aβ generation as well as increase intermediate APP cleavage products that exacerbate pathological changes.

The first isogenic AD model was an allelic series *PSEN1*$^{\Delta E9}$ mutant knock-in and *PSEN1*$^{WT/KO}$ generated via transcription activator-like effector nucleases [54]. The *ΔE9* mutation refers to loss of exon 9 expression, which prevents endoproteolysis of PSEN1 and its subsequent full enzymatic activation. *PSEN1*$^{\Delta E9}$ mutant neurons had increased Aβ42:Aβ40 ratio (from reduced Aβ40) as well as increased APP-CTFs (C99 and C83), which was not seen in *PSEN1*$^{WT/KO}$, supporting a GOF rather than LOF role for the mutation. Although this study is consistent with elevated

C99 driving pathogenesis, *PSEN1*$^{\Delta E9}$ mutant neurons did not have tau phenotypes, which has been consistent for *PSEN1* mutant hiPSC models from several groups for unknown reasons. A much larger panel of isogenic FAD lines generated by CRISPR-mediated knock-in was generated with a focus on early endosome morphology and global gene expression [55]. This included APP mutants (Swedish, A692G, V717G), PSEN1 mutants (M146V, L166P, M233L), and APP/PSEN1 compound mutants generated by sequential rounds of CRISPR-editing. Enlarged early endosomes were seen in mutant neuron models in a manner driven by C99 rather than Aβ, as demonstrated with similar β- and γ-secretase inhibitor strategies described as above.

These studies suggest that C99 is a major driver of AD and is further supported by a recent report which shows that C99 levels (as measured via proximity ligation assays) correlate much better with AD disease progression than Aβ in human brain samples [56]. This leads to additional questions such as: What is the mechanism(s) of how C99 drives AD-related pathology? Why is this not seen with C83, the intermediate cleavage product produced by α-cleavage of APP?

One functional difference between C99 and C83 is the presence of a cholesterol binding domain in the former [57]. This was recently the focus of a report my lab contributed to, which argued that C99 serves as a cholesterol sensor in the endoplasmic reticulum (ER) in a specialized structure called the mitochondria-associated ER membrane (MAM) [58]. Increase of wildtype C99 expression, but not a mutant deficient in cholesterol binding, downregulated de novo synthesis of cholesterol, promoted internalization of extracellular cholesterol, increased cholesterol trafficking from the plasma membrane to the ER (MAM), increased the size of the MAM to activate sphingomyelinases to expose shielded cholesterol found in lipid rafts, and increased cholesterol esterification by acyl-coenzyme A:cholesterol acyltransferase 1. Excess cholesterol is detoxified when converted to cholesterol esters which is also the primary component of lipid droplets seen by Alois Alzheimer in his initial observations, as alluded to earlier [4]. Although much of the work was done in other cellular systems, some of these effects on cholesterol were also modeled in undifferentiated hiPSCs in which the *APP*^{London} mutation was knocked into both alleles as well as the isogenic parent line. Genetic risk factors for AD that promote C99 accumulation would drive this cholesterol dysregulation and create a feed-forward loop of continuous cholesterol internalization and conversion to CEs.

Cholesterol has long been thought to play an important role in AD. Multiple genetic risk factors for LOAD are associated with cholesterol such as APOE and clusterin [59]. Increasing cholesterol via exogenous addition has been shown to mimic increased pTau, and enlarged early endosome phenotypes seen in AD (as well as increase Aβ) [60,61]; AD pathology can also be seen in vivo in mice via shRNA-mediated knockdown of Cyp46A1 (enzyme which converts cholesterol to 24-hydroxycholesterol) [62], and can affect pTau and Aβ levels in hiPSC-derived neurons [63]. I propose that dysregulation of cholesterol unifies both FAD and LOAD and serves as a central hub to drive AD-related phenotypes in isogenic hiPSC AD models generated via CRISPR (Fig. 6.2).

Concerns about deleterious on-target effects by CRISPR in ADRD models

Potential off-target effects of CRISPR-mediated editing typically have been considered by researchers and assessed by methods such as Sanger sequencing of predicted off-targets and whole exome/genome sequencing, among others. However, potential deleterious on-target effects of genome editing has also recently arisen as a significant concern [64–66].

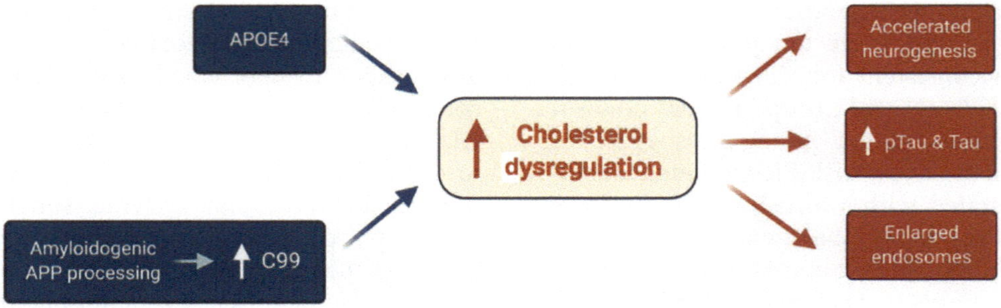

FIGURE 6.2 Model of interrelated AD phenotypes. *Source: Courtesy of Andrew Sproul and Saera Song.*

This has also been shown to be true for HDR-mediated CRISPR-edited hiPSC lines, in which supposed homozygous knock-in lines were demonstrated to be hemizygous with deletions too large to be picked up by conventional HDR Sanger sequencing screening methods [67]. In addition, some clones also showed significant loss of heterozygosity of adjacent single nucleotide polymorphisms (SNPs) in the targeted allele. This could have profound effects on disease modeling including for ADRDs. For instance, hemizygous $APP^{Swedish}$ *(KM670/671NL)* mutant neurons produce ~50% less Aβ than isogenic $APP^{Swedish}$ homozygous mutant neurons [67]. Thus strategies to address potential deleterious on-target effects should be incorporated in gene-edited lines for disease modeling, such as quantitative genotyping polymerase chain reaction of amplified gDNA from the targeted region to assess allelic copy number [67]. CRISPR genome-editing techniques that do not rely on double-strand breaks (DSBs) may alleviate this problem, as has been suggested for base editors [68].

CRISPR activation/inhibition/knock-out for modeling and genetic screens

Although isogenic ADRD disease models allow one to get deep insight into a particular genetic risk variant, CRISPR technologies can be used more broadly to interrogate many genes through either genome-wide or targeted candidate pool screens. Large-scale genome-wide association studies have implicated hundreds of loci in the pathogenesis of LOAD including a plethora of intergenic SNPs which could act as distal enhancers [69]. In order to get an initial handle on how these associated genetic risk factors, or conversely genetic protective factors, might be affecting a given ADRD-related phenotype, CRISPR-mediated screens can greatly increase the throughput. Likewise, previously unknown genetic modifiers of an ADRD-relevant cellular behavior identified through unbiased screens might identify novel therapeutic targets. CRISPR screens could include LOF via indel induction generated by nonhomologous end joining after introduction of a DSB, LOF through CRISPRi, or GOF via CRISPRa. Note that of course CRISPRi and CRISPRa can be used to look at individual disease-relevant targets to replace or complement strategies such as small interfering RNA (siRNA)-mediated knockdown or plasmid/viral overexpression respectively, although this does not unleash their full potential.

CRISPR activation/inhibiton for individual targets

The first study to use CRISPRa in the context of AD was to increase APP and BACE1 levels in human PSEN1/2 mutant and control fibroblasts [70]. This demonstrated reduced carboxypeptidase activity of the γ-secretase complex toward APP, which means the processivity of multiple cuts of APP was reduced to produce longer more oligomerogenic forms of Aβ. The first report using CRISPRi in the context of AD was in a study demonstrating that treatment of human hESC-derived neurons with recombinant APOE drives APP transcription via a novel noncanonical DLK-MKK7-ERK pathway, and that APOE4 (risk) did so more than APOE3 (neutral), which did so more than APOE2 (protective) [71]. CRISPRi was used along with conventional siRNAs to interrogate nodes of the noncanonical signaling cascade identified. Elevation of APP transcription has significant potential consequences, as rare APP duplication has been reported to be sufficient to cause EOAD, and three copies of APP on chromosome 21 in individuals with Down syndrome (DS) is thought to underlie the high propensity of dementia in that population [72,73]. However, another study using CRISPRa in human hESC to increase APP levels and DS hiPSCs which were normalized to two copies of APP via CRISPR, suggests that some phenotypes seen in DS hiPSC-derived neurons such as tau hyperphosphorylation are APP copy number-independent [74]. This supports the argument that additional genes on chromosome 21 are critical to promote AD-like dementia in DS. Interestingly, EOAD APP duplication mutations in fact include a few other proximal genes as well as theoretical enhancer elements in noncoding regions [72].

CRISPR activation/inhibition/knock-out for genetic screens

Only a few studies have been published to date using CRISPR genetic screens specifically in the context of ADRDs, although we and many other investigators are currently performing such screens. In the context of directly assessing effectors of AD neuropathology, a global CRISPR knock-out screen in the Neuro2a cell line was performed with Aβ levels in the media used as a readout [75]. Calcium and integrin-binding protein 1 was identified as a negative regulator of Aβ production, potentially through control of the subcellular localization of the γ-secretase component nicastrin. In a somewhat similar vein, another recent candidate-based siRNA screen in human hiPSC-derived neurons and commercial human fetal astrocytes identified multiple regulators of Aβ production, and a top hit FERM2 which affected both Aβ42 levels and tau phosphorylation was confirmed via CRISPR-mediated knock-out [76]. The mechanism of how FERM2, a scaffold protein with known functions with integrin signaling, actin skeleton, and cell shape, affects AD remains to be elucidated.

Another CRISPRi screen focused on regulators of tau pathology, as misfolded tau has been demonstrated to travel between synaptically connected neurons to convert normal folded tau to a pathological conformation, conceptionally similar to the prion protein in Creutzfeldt–Jacob disease [77]. HEK293T cells were used to screen for genes that promoted tau aggregation using a fluorescence resonance energy transfer-based assay, and endosomal sorting complexes required for transport pathway members such as CHMP6 and CHMP2A/CHMP2B were identified. Additional relevant CRISPR screening approaches have focused on immune cell function, which may have relevance to microglial roles in ADRDs. These included CRISPR knock-out screens to identify modulators of NLRP3 inflammasome activation in immortalized mouse macrophages [78] and regulators of phagocytosis in the human myeloid cell line U937 [79]. Although these initial CRISPR screens have generated some intriguing findings and genetic targets for further study, we are at the tip of the iceberg of these powerful approaches being applied robustly to hiPSC-derived ADRD-relevant cell types.

CRISPR for disease therapeutics

In addition to using CRISPR technologies for disease modeling to better understand ADRD pathogenesis, these techniques may be applied more directly for therapeutic purposes. This includes gene therapy approaches targeting key ADRD genetic factors utilizing CRISPR-mediated gene editing and gene correction of mutations/introduction of transgenes in patient hiPSCs prior to regenerative medicine approaches.

CRISPR-mediated editing as gene therapy for ADRD

Although gene therapy using CRISPR to target CNS cells affected in ADRDs has significant technical and safety hurdles to overcome before appearing in the clinic, work has begun to develop proofs of principle for this approach. In the context of ADRD, the following strategies have emerged: (1) specifically targeting and disrupting mutant versions of alleles or correcting them by HDR-mediated mechanism; (2) targeting modification of disease-relevant alleles in a manner to maintain at least some protein functionality; and (3) complete knock-out of a disease-relevant allele. hiPSC models have been used to validate some of these approaches.

As described earlier for some studies, a variety of ADRD pathogenic mutations/risk alleles have been gene corrected to control in hiPSCs, including tau (A152T, N279K) [18,80], PSEN1 (ΔE9) [81], PSEN2 (N141I) [82], and APOE4 (to APOE3) [23,26,83]. Although these were primarily developed for disease modeling purposes, if efficient CRISPR HDR-mediated gene editing is developed, theoretically similar gene-correction strategies could be used clinically. A more practical approach is to use CRISPR for nonhomologous end joining (NHEJ)-mediated alteration or complete disruption of disease-relevant genes. The first study to try this approach used a single guide RNA (sgRNA) to target the $APP^{Swedish}$ mutant allele in human fibroblasts and in vivo in an $APP^{Swedish}$ transgenic model using adeno-associated viruses (AAV) delivery [84]. The $APP^{Swedish}$ mutation is sufficient to cause FAD [85], and historically has been used in many mouse FAD models. There was limited engagement with the target in vivo (1.3% of transgenic alleles showed $APP^{Swedish}$ indels). This may have resulted at least in part from greater than 100 copies of the mutant allele in this model.

The second study, which my own lab collaborated on, decided instead to target nonmutant APP, but in a manner that could potentially maintain some APP functions while reducing its ability to produce toxic proteolytic fragments [86]. The distal C-terminus of APP was targeted with a sgRNA, which when edited via NHEJ to introduce indels, reduced β-cleavage of APP and resulting toxic C99 and Aβ proteolytic fragments while promoting beneficial α-cleavage of APP. This was demonstrated in human neurons derived from isogenic control and CRISPR-knock-in APP^{London} (V717I) mutant hiPSCs, as well as robustly in vivo in wild type mice. This likely resulted from removal of the YENPTY domain responsible for APP endosomal trafficking where β-cleavage and amyloidogenic processing occurs, as opposed to the plasma membrane which is the primary subcellular localization for α-cleavage. As this was the last coding exon, truncated transcripts could avoid nonsense-mediated decay. One advantage of targeting APP more generally rather than a specific mutation is that all EOAD cases (at least familial forms) are thought to be dependent on altered amyloidogenic processing of APP to cause AD, and as such should be amenable to this approach. This includes over 185 EOAD mutations identified in PSEN1 [87], the most common gene affected in EOAD and an unattractive direct target for CRISPR-mediated knock-out due to it being the catalytic core of the γ-secretase complex which cleaves Notch among other targets.

A third approach is to use CRISPR to completely knock-out nonmutated versions of key pathogenic genes, such as was done for BACE1, the enzyme responsible amyloidogenic β-cleavage of APP, using amphipathic nanocomplexes containing Cas9-sgRNA ribonucleoproteins [88]. This approach was done successfully in vivo in two FAD mouse models, where significant reduction in BACE1 in turn reduced amyloid burden and partially rescued cognitive deficits. Although BACE1 itself may not be a good target—given BACE1's capacity to cleave multiple unintended substrates and the recent failures of BACE1 inhibitors in the clinic [89]—this study validates nanocomplexes as a potential delivery mechanism for CRISPR-mediated therapeutics targeting neurons in the CNS. Another potential ADRC-specific knock-out target is currently the focus of an industrial partnership between Sangamo and Biogen to develop knock-out of tau via zinc fingers as a potential therapeutic for AD and other tauopathies such as FTD [90].

In terms of the potential practicality of application of CRISPR as a direct therapeutic, targeting familial EOAD would be the most tractable, as many EOAD mutations have 100% penetrance and, in some cases, the range of the age of onset has been well described. For those expected to get EOAD in their 40s and 50s, gene therapy might be well worth the risk. For mitigation of Aβ/C99, targeting the C-terminus of APP would greatly broaden the scope of a single therapeutic agent as compared to mutant allele-specific approaches. However, as amyloid has been shown to build up decades before symptoms [91], the therapeutic window might close significantly before disease onset. Targeting tau, which builds up later during the disease, might allow this to shift to later time points, and would be desirable as an approach for autosomal-dominant FTD.

Using CRISPR as a therapeutic in LOAD patients would be much less likely to appear in the clinic, as no genetic factor is sufficient to cause the disease, making definitive prognosis challenging. Potentially significant improvements in AD biomarkers and polygenic risk score assessment (from patient whole genome sequencing data) might allow one to have a high degree of diagnostic confidence of the likelihood of LOAD. However, whether this would be sufficient to motivate patients to seek such an invasive treatment a decade or more before symptoms and whether insurance companies would cover such a procedure remains questionable.

CRISPR-mediated gene correction or transgene insertion for cell replacement therapies for ADRDs

Excitingly, hPSC-derived cell replacement therapies are in clinical trials now for a variety of disorders, including those affecting the eyes as described elsewhere in this book. The first neurodegenerative disease cell replacement therapy to have received approvals for clinical trials is Parkinson's disease. The first trial began in Japan in 2018 using allogenic hiPSCs, followed by studies which started in China and Australia using hESCs [92]. The first US trial was approved in early 2021 using hESCs as a starting material [93]. Parkinson's disease is a challenging but easier target for cell replacement therapy than ADRDs, as it involves a reasonable well-defined neural circuit with specific subclasses of dopaminergic neurons being primarily affected. The approach is also supported scientifically by previous studies which demonstrated long-term benefits from human fetal dopaminergic neuron transplants [94–96].

ADRDs do have some initial areas of selective vulnerability, such as the entorhinal cortex and hippocampus, which are stereotypically affected early in AD [97]. As the disease progresses much larger areas of the cortex are affected. Targeted cell replacement using hiPSC-derived specific subclasses of hippocampal neurons, namely CA1 and CA3 glutamatergic projection neurons, could provide benefit; however, hiPSC cortical differentiation protocols have not reached this level of anatomical precision to date. Another potential hiPSC-derived cell type that could be used for regenerative medicine purposes would be basal forebrain cholinergic neurons (BFCNs), another vulnerable population in AD. The cholinergic hypothesis of AD [98], a predecessor to the amyloid hypothesis, postulates that loss of BFCNs and associated cholinergic signaling was a key driver of cognitive deficits in AD, similar conceptually to loss of dopaminergic signaling in Parkinson's disease. The importance of BFCNs is underscored by the use of cholinesterase inhibitors to treat AD, which increase the levels of acetylcholine by preventing its breakdown in the synaptic cleft; this is one of the only two classes of therapeutics approved for treatment of AD symptoms. In support of this approach, hESC-derived BFCNs improved cognitive function in two AD mouse models [99].

For most hiPSC-derived cell replacement strategies including those for neurodegenerative diseases, a single hPSC line or a small pool of allogenic hiPSCs will be used initially as a source for differentiated cells for both practical and economic reasons. In the long run, personalized medicine approaches using a patient's own hiPSCs for transplantation will likely be used once technically and economically feasible. In this case, it would be prudent to use CRISPR-mediated gene correction to remove strong ADRD-related mutations such as those that cause FAD or FTD in patient hiPSCs before differentiation and transplantation, to ensure that these neurons would remain functional for decades posttransplantation in these younger subjects with familial forms of ADRDs.

A potentially closer-term application for CRISPR technologies to aid regenerative therapeutics would be to knock-in transgenes that could provide clinical benefit when overexpressed in a cellular therapeutic. Transplantation of either mouse or primary human neural stem cells (NSCs) improved cognition and synaptic growth in AD mouse models [100,101], likely primarily through trophic support mechanisms such as BDNF release rather than via cell replacement or directly affecting amyloid or tau pathology. One could likely substitute hiPSC-derived NPCs for fetal NSCs, and in the long term potentially autologous hiPSC-derived NPCs. These could be further modified to express additional beneficial transgenes by safely introducing them to a safe harbor locus such as AAVS1 using CRISPR-mediated knock-in. For instance, mouse NSCs overexpressing a secreted form of neprilysin, an enzyme which degrades Aβ, also reduced neuropathology in AD mouse models [102]. Neurotrophic factors, immune modulators, cholesterol modulators, microRNAs such as those that target tau, or other beneficial expression constructs could be introduced in a similar fashion via CRISPR technologies to make a multifunctional cellular therapy.

Future directions

CRISPR and hiPSC technologies have already aided us in helping to better understand ADRDs, as well as potentially being used as direct therapeutics. Although this chapter focused primarily on AD, similar approaches are being utilized for other dementias and neurodegenerative diseases. CRISPR-mediated gene editing to create isogenic systems is now the standard when possible to do so in cellular models of disease. Although early studies using CRISPR screens—knock-out, inhibition, and activation—have had limited use in ADRDs, particularly in hiPSC models, their full power to interrogate genome-wide effects on disease-relevant phenotypes will revolutionize our understanding of ADRD pathogenesis. One area where this will hold true is in the improved mechanistic understanding of the hundreds of SNPs that either increase or reduce ADRD risk. This may lead to more of a precision medicine approach to dementia where, for example, LOAD is divided into different subvariants in a spectrum rather than a single disorder.

Dementia continues to incite fear worldwide, and the failure of most clinical trials designed to combat these diseases reduces the hope of a cure with current disease intervention strategies. However, given the significant research funding and resources dedicated to dementia research, techniques involving CRISPR and hiPSC technologies offer hope and encouragement for a future therapeutic solution.

References

[1] R. Brookmeyer, et al., Forecasting the global burden of Alzheimer's disease, Alzheimers Dement. 3 (3) (2007) 186–191.

[2] A. Wimo, et al., The worldwide costs of dementia 2015 and comparisons with 2010, Alzheimers Dement. 13 (1) (2017) 1–7.

[3] J.L. Cummings, T. Morstorf, K. Zhong, Alzheimer's disease drug-development pipeline: few candidates, frequent failures, Alzheimers Res. Ther. 6 (4) (2014) 37.

[4] P. Foley, Lipids in Alzheimer's disease: a century-old story, Biochim. Biophys. Acta 1801 (8) (2010) 750–753.

[5] R. Cacace, K. Sleegers, C. Van, Broeckhoven, Molecular genetics of early-onset Alzheimer's disease revisited., Alzheimers Dement. 12 (6) (2016) 733–748.

[6] M.L. Cuccaro, et al., SORL1 mutations in early- and late-onset Alzheimer disease, Neurol. Genet. 2 (6) (2016) e116.

[7] A.C. Naj, G.D. Schellenberg, Alzheimer's Disease Genetics Consortium, Genomic variants, genes, and pathways of Alzheimer's disease: an overview., Am. J. Med. Genet. B Neuropsychiatr. Genet. 174 (1) (2017) 5–26.

[8] C.M. Wolfe, et al., The Role of APOE and TREM2 in Alzheimer's disease-current understanding and perspectives, Int. J. Mol. Sci. 20 (1) (2018) 81.

[9] R. Sengoku, Aging and Alzheimer's disease pathology, Neuropathology 40 (1) (2020) 22–29.

[10] D. Cheng, et al., Type 2 diabetes and late-onset Alzheimer's disease, Dement. Geriatr. Cogn. Disord. 31 (6) (2011) 424–430.

[11] L.G. Exalto, et al., Midlife risk score for the prediction of dementia four decades later, Alzheimers Dement. 10 (5) (2014) 562–570.

[12] G. Tosto, et al., The role of cardiovascular risk factors and stroke in familial Alzheimer disease, JAMA Neurol. 73 (10) (2016) 1231–1237.

[13] D.J. Selkoe, J. Hardy, The amyloid hypothesis of Alzheimer's disease at 25 years, EMBO Mol. Med. 8 (6) (2016) 595–608.

[14] T. Jonsson, et al., A mutation in APP protects against Alzheimer's disease and age-related cognitive decline, Nature 488 (7409) (2012) 96–99.

[15] S.S. Kwak, et al., Amyloid-beta42/40 ratio drives tau pathology in 3D human neural cell culture models of Alzheimer's disease, Nat. Commun. 11 (1) (2020) 1377.

[16] F.M. Elahi, B.L. Miller, A clinicopathological approach to the diagnosis of dementia, Nat. Rev. Neurol. 13 (8) (2017) 457–476.

[17] A.A. Sproul, Being human: the role of pluripotent stem cells in regenerative medicine and humanizing Alzheimer's disease models, Mol. Asp. Med. 43–44 (2015) 54–65.

[18] H. Fong, et al., Genetic correction of tauopathy phenotypes in neurons derived from human induced pluripotent stem cells, Stem Cell Rep. 1 (3) (2013) 226–234.

[19] D. Paquet, et al., Efficient introduction of specific homozygous and heterozygous mutations using CRISPR/Cas9, Nature 533 (7601) (2016) 125–129.

[20] L.A. Farrer, et al., Effects of age, sex, and ethnicity on the association between apolipoprotein E genotype and Alzheimer disease. A meta-analysis. APOE and Alzheimer Disease Meta Analysis Consortium, JAMA 278 (16) (1997) 1349–1356.

[21] J.F. Arboleda-Velasquez, et al., Resistance to autosomal dominant Alzheimer's disease in an APOE3 Christchurch homozygote: a case report, Nat. Med. 25 (11) (2019) 1680–1683.

[22] J. Kim, J.M. Basak, D.M. Holtzman, The role of apolipoprotein E in Alzheimer's disease, Neuron 63 (3) (2009) 287–303.

[23] Y.T. Lin, et al., APOE4 causes widespread molecular and cellular alterations associated with Alzheimer's disease phenotypes in human iPSC-derived brain cell types, Neuron 98 (6) (2018) 1141–1154. e7.

[24] P.T. Xu, et al., Specific regional transcription of apolipoprotein E in human brain neurons, Am. J. Pathol. 154 (2) (1999) 601–611.

[25] A.M. Cataldo, et al., Increased neuronal endocytosis and protease delivery to early endosomes in sporadic Alzheimer's disease: neuropathologic evidence for a

mechanism of increased beta-amyloidogenesis, J. Neurosci. 17 (16) (1997) 6142–6151.
[26] C. Wang, et al., Gain of toxic apolipoprotein E4 effects in human iPSC-derived neurons is ameliorated by a small-molecule structure corrector, Nat. Med. 24 (5) (2018) 647–657.
[27] A. Knupp, et al., Depletion of the AD risk gene SORL1 selectively impairs neuronal endosomal traffic independent of amyloidogenic APP processing, Cell Rep. 31 (9) (2020) 107719.
[28] M. Olah, et al., A transcriptomic atlas of aged human microglia, Nat. Commun. 9 (1) (2018) 539.
[29] J.W. Blanchard, et al., Reconstruction of the human blood-brain barrier in vitro reveals a pathogenic mechanism of APOE4 in pericytes, Nat. Med. 26 (6) (2020) 952–963.
[30] M. Boldrini, et al., Human hippocampal neurogenesis persists throughout aging, Cell Stem Cell 22 (4) (2018) 589–599. e5.
[31] E.P. Moreno-Jimenez, et al., Adult hippocampal neurogenesis is abundant in neurologically healthy subjects and drops sharply in patients with Alzheimer's disease, Nat. Med. 25 (4) (2019) 554–560.
[32] M.K. Tobin, et al., Human hippocampal neurogenesis persists in aged adults and Alzheimer's disease patients, Cell Stem Cell 24 (6) (2019) 974–982. e3.
[33] P. He, Y. Shen, Interruption of beta-catenin signaling reduces neurogenesis in Alzheimer's disease, J. Neurosci. 29 (20) (2009) 6545–6557.
[34] K. Jin, et al., Enhanced neurogenesis in Alzheimer's disease transgenic (PDGF-APPSw,Ind) mice, Proc. Natl Acad. Sci. U.S.A. 101 (36) (2004) 13363–13367.
[35] K. Boekhoorn, M. Joels, P.J. Lucassen, Increased proliferation reflects glial and vascular-associated changes, but not neurogenesis in the presenile Alzheimer hippocampus, Neurobiol. Dis. 24 (1) (2006) 1–14.
[36] Y. Mu, F.H. Gage, Adult hippocampal neurogenesis and its role in Alzheimer's disease, Mol. Neurodegener. 6 (2011) 85.
[37] M.A. Lopez-Toledano, M.L. Shelanski, Increased neurogenesis in young transgenic mice overexpressing human APP(Sw, Ind), J. Alzheimers Dis. 12 (3) (2007) 229–240.
[38] C.P. Yang, et al., ApoE is required for maintenance of the dentate gyrus neural progenitor pool, Development 138 (20) (2011) 4351–4362.
[39] Y. Tensaouti, et al., ApoE regulates the development of adult newborn hippocampal neurons, eNeuro 5 (2018) 4. ENEURO.0155–18.2018.
[40] M.A. Lopez-Toledano, M.L. Shelanski, Neurogenic effect of beta-amyloid peptide in the development of neural stem cells, J. Neurosci. 24 (23) (2004) 5439–5444.
[41] Q.H. Ma, et al., A TAG1-APP signalling pathway through Fe65 negatively modulates neurogenesis, Nat. Cell Biol. 10 (3) (2008) 283–294.
[42] K. Ghosal, A. Stathopoulos, S.W. Pimplikar, APP intracellular domain impairs adult neurogenesis in transgenic mice by inducing neuroinflammation, PLoS One 5 (7) (2010) e11866.
[43] A.A. Sproul, et al., Characterization and molecular profiling of PSEN1 familial Alzheimer's disease iPSC-derived neural progenitors, PLoS One 9 (1) (2014) e84547.
[44] R. Sherrington, et al., Cloning of a gene bearing missense mutations in early-onset familial Alzheimer's disease, Nature 375 (6534) (1995) 754–760.
[45] K. Meyer, et al., REST and neural gene network dysregulation in iPSC models of Alzheimer's disease, Cell Rep. 26 (5) (2019) 1112–1127. e9.
[46] N. Ballas, G. Mandel, The many faces of REST oversee epigenetic programming of neuronal genes, Curr. Opin. Neurobiol. 15 (5) (2005) 500–506.
[47] T. Lu, et al., REST and stress resistance in ageing and Alzheimer's disease, Nature 507 (7493) (2014) 448–454.
[48] C. Arber, et al., Familial Alzheimer's disease mutations in PSEN1 lead to premature human stem cell neurogenesis, Cell Rep. 34 (2) (2021) 108615.
[49] J. Yang, et al., Early pathogenic event of Alzheimer's disease documented in iPSCs from patients with PSEN1 mutations, Oncotarget 8 (5) (2017) 7900–7913.
[50] E. Vico Varela, G. Etter, S. Williams, Excitatory-inhibitory imbalance in Alzheimer's disease and therapeutic significance, Neurobiol. Dis. 127 (2019) 605–615.
[51] S. Moore, et al., APP metabolism regulates tau proteostasis in human cerebral cortex neurons, Cell Rep. 11 (5) (2015) 689–696.
[52] M.A. Israel, et al., Probing sporadic and familial Alzheimer's disease using induced pluripotent stem cells, Nature 482 (7384) (2012) 216–220.
[53] B.P. Imbimbo, G.A. Giardina, -secretase inhibitors and modulators for the treatment of Alzheimer's disease: disappointments and hopes, Curr. Top. Med. Chem. 11 (12) (2011) 1555–1570.
[54] G. Woodruff, et al., The presenilin-1 DeltaE9 mutation results in reduced gamma-secretase activity, but not total loss of PS1 function, in isogenic human stem cells, Cell Rep. 5 (4) (2013) 974–985.
[55] D. Kwart, et al., A large panel of isogenic APP and PSEN1 mutant human iPSC neurons reveals shared endosomal abnormalities mediated by APP beta-CTFs, Not Abeta, Neuron 104 (2) (2019) 256–270. e5.
[56] M.V. Pulina, et al., C99 selectively accumulates in vulnerable neurons in Alzheimer's disease, Alzheimers Dement. 16 (2) (2020) 273–282.

[57] P.J. Barrett, et al., The amyloid precursor protein has a flexible transmembrane domain and binds cholesterol, Science 336 (6085) (2012) 1168–1171.

[58] J. Montesinos, et al., The Alzheimer's disease-associated C99 fragment of APP regulates cellular cholesterol trafficking, EMBO J. 39 (20) (2020) e103791.

[59] H.K. Dong, et al., Integrated late onset Alzheimer's disease (LOAD) susceptibility genes: cholesterol metabolism and trafficking perspectives, Gene 597 (2017) 10–16.

[60] C. Marquer, et al., Increasing membrane cholesterol of neurons in culture recapitulates Alzheimer's disease early phenotypes, Mol. Neurodegener. 9 (2014) 60.

[61] C. Marquer, et al., Local cholesterol increase triggers amyloid precursor protein-Bace1 clustering in lipid rafts and rapid endocytosis, FASEB J. 25 (4) (2011) 1295–1305.

[62] F. Djelti, et al., CYP46A1 inhibition, brain cholesterol accumulation and neurodegeneration pave the way for Alzheimer's disease, Brain 138 (Pt 8) (2015) 2383–2398.

[63] R. van der Kant, et al., Cholesterol metabolism is a druggable axis that independently regulates tau and amyloid-beta in iPSC-derived Alzheimer's disease neurons, Cell Stem Cell 24 (3) (2019) 363–375. e9.

[64] M.V. Zuccaro, et al., Allele-specific chromosome removal after Cas9 cleavage in human embryos, Cell 183 (6) (2020) 1650–1664. e15.

[65] M. Kosicki, K. Tomberg, A. Bradley, Repair of double-strand breaks induced by CRISPR-Cas9 leads to large deletions and complex rearrangements, Nat. Biotechnol. 36 (8) (2018) 765–771.

[66] F. Adikusuma, et al., Large deletions induced by Cas9 cleavage, Nature 560 (7717) (2018) E8–E9.

[67] I. Weisheit, et al., Detection of deleterious on-target effects after HDR-mediated CRISPR editing, Cell Rep. 31 (8) (2020) 107689.

[68] Y. Song, et al., Large-fragment deletions induced by Cas9 cleavage while not in the BEs system, Mol. Ther. Nucleic Acids 21 (2020) 523–526.

[69] B.W. Kunkle, et al., Genetic meta-analysis of diagnosed Alzheimer's disease identifies new risk loci and implicates Abeta, tau, immunity and lipid processing, Nat. Genet. 51 (3) (2019) 414–430.

[70] K. Inoue, L.M.A. Oliveira, A. Abeliovich, CRISPR transcriptional activation analysis unmasks an occult gamma-secretase processivity defect in familial Alzheimer's disease skin fibroblasts, Cell Rep. 21 (7) (2017) 1727–1736.

[71] Y.A. Huang, et al., ApoE2, ApoE3, and ApoE4 differentially stimulate APP transcription and Abeta secretion, Cell 168 (3) (2017) 427–441. e21.

[72] A. Rovelet-Lecrux, et al., APP locus duplication causes autosomal dominant early-onset Alzheimer disease with cerebral amyloid angiopathy, Nat. Genet. 38 (1) (2006) 24–26.

[73] B. Decourt, et al., Recent perspectives on APP, secretases, endosomal pathways and how they influence Alzheimer's related pathological changes in Down syndrome, J. Alzheimer. Dis. Parkinsonism Suppl 7 (2013) 002.

[74] D.A. Ovchinnikov, et al., The impact of APP on Alzheimer-like pathogenesis and gene expression in Down syndrome iPSC-derived neurons, Stem Cell Rep. 11 (1) (2018) 32–42.

[75] Y.W. Chiu, et al., Identification of calcium and integrin-binding protein 1 as a novel regulator of production of amyloid beta peptide using CRISPR/Cas9-based screening system, FASEB J. 34 (6) (2020) 7661–7674.

[76] S.E. Sullivan, et al., Candidate-based screening via gene modulation in human neurons and astrocytes implicates FERMT2 in Abeta and TAU proteostasis, Hum. Mol. Genet. 28 (5) (2019) 718–735.

[77] J.J. Chen, et al., Compromised function of the ESCRT pathway promotes endolysosomal escape of tau seeds and propagation of tau aggregation, J. Biol. Chem. 294 (50) (2019) 18952–18966.

[78] J.L. Schmid-Burgk, et al., A genome-wide CRISPR (clustered regularly interspaced short palindromic repeats) screen identifies NEK7 as an essential component of NLRP3 inflammasome activation, J. Biol. Chem. 291 (1) (2016) 103–109.

[79] M.S. Haney, et al., Identification of phagocytosis regulators using magnetic genome-wide CRISPR screens, Nat. Genet. 50 (12) (2018) 1716–1727.

[80] A.L. Hallmann, et al., Astrocyte pathology in a human neural stem cell model of frontotemporal dementia caused by mutant TAU protein, Sci. Rep. 7 (2017) 42991.

[81] M. Oksanen, et al., PSEN1 mutant iPSC-derived model reveals severe astrocyte pathology in Alzheimer's disease, Stem Cell Rep. 9 (6) (2017) 1885–1897.

[82] M. Ortiz-Virumbrales, et al., CRISPR/Cas9-correctable mutation-related molecular and physiological phenotypes in iPSC-derived Alzheimer's PSEN2 (N141I) neurons, Acta Neuropathol. Commun. 5 (1) (2017) 77.

[83] A.R. Wadhwani, et al., Neuronal apolipoprotein E4 increases cell death and phosphorylated tau release in alzheimer disease, Ann. Neurol. 85 (5) (2019) 726–739.

[84] B. Gyorgy, et al., CRISPR/Cas9 mediated disruption of the Swedish APP allele as a therapeutic approach for early-onset Alzheimer's disease, Mol. Ther. Nucleic Acids 11 (2018) 429–440.

[85] C. Haass, et al., The Swedish mutation causes early-onset Alzheimer's disease by beta-secretase cleavage within the secretory pathway, Nat. Med. 1 (12) (1995) 1291–1296.

[86] J. Sun, et al., CRISPR/Cas9 editing of APP C-terminus attenuates beta-cleavage and promotes alpha-cleavage, Nat. Commun. 10 (1) (2019) 53.
[87] C.M. Karch, C. Cruchaga, A.M. Goate, Alzheimer's disease genetics: from the bench to the clinic, Neuron 83 (1) (2014) 11–26.
[88] H. Park, et al., In vivo neuronal gene editing via CRISPR-Cas9 amphiphilic nanocomplexes alleviates deficits in mouse models of Alzheimer's disease, Nat. Neurosci. 22 (4) (2019) 524–528.
[89] B. Das, R. Yan, A close look at BACE1 inhibitors for Alzheimer's disease treatment, CNS Drugs 33 (3) (2019) 251–263.
[90] T. Brady, A new approach for Alzheimer's, 2020; Available from: https://www.bio.org/blogs/new-approach-alzheimers.
[91] J.M. Long, D.M. Holtzman, Alzheimer disease: an update on pathobiology and treatment strategies, Cell 179 (2) (2019) 312–339.
[92] T.B. Stoker, R.A. Barker, Recent developments in the treatment of Parkinson's disease, F1000Res 9 (2020) F1000.
[93] J. Carvalho, Cell therapy DA01 cleared for phase 1 trial in advanced disease patients, 2021; Available from: https://parkinsonsnewstoday.com/2021/01/11/fda-clears-phase-1-trial-da01-bluerock-cell-therapy-advanced-parkinsons/.
[94] Z. Kefalopoulou, et al., Long-term clinical outcome of fetal cell transplantation for Parkinson disease: two case reports, JAMA Neurol. 71 (1) (2014) 83–87.
[95] P.J. Hallett, et al., Long-term health of dopaminergic neuron transplants in Parkinson's disease patients, Cell Rep. 7 (6) (2014) 1755–1761.
[96] W. Li, et al., Extensive graft-derived dopaminergic innervation is maintained 24 years after transplantation in the degenerating parkinsonian brain, Proc. Natl Acad. Sci. U.S.A. 113 (23) (2016) 6544–6549.
[97] B.T. Hyman, et al., Alzheimer's disease: cell-specific pathology isolates the hippocampal formation, Science 225 (4667) (1984) 1168–1170.
[98] H. Hampel, et al., The cholinergic system in the pathophysiology and treatment of Alzheimer's disease, Brain 141 (7) (2018) 1917–1933.
[99] W. Yue, et al., ESC-derived basal forebrain cholinergic neurons ameliorate the cognitive symptoms associated with Alzheimer's disease in mouse models, Stem Cell Rep. 5 (5) (2015) 776–790.
[100] M. Blurton-Jones, et al., Neural stem cells improve cognition via BDNF in a transgenic model of Alzheimer disease, Proc. Natl Acad. Sci. U.S.A. 106 (32) (2009) 13594–13599.
[101] R.R. Ager, et al., Human neural stem cells improve cognition and promote synaptic growth in two complementary transgenic models of Alzheimer's disease and neuronal loss, Hippocampus 25 (7) (2015) 813–826.
[102] M. Blurton-Jones, et al., Neural stem cells genetically-modified to express neprilysin reduce pathology in Alzheimer transgenic models, Stem Cell Res. Ther. 5 (2) (2014) 46.

CHAPTER

7

Gene editing for the cornea

M. Andrew Nesbit and C.B. Tara Moore

Biomedical Sciences Research Institute, Ulster University, Coleraine, United Kingdom

The cornea

The cornea is an avascular, transparent tissue forming the outermost part of the anterior segment of the eye. Its two main functions are to act as a protective barrier for the other components of the eye and to provide around two-thirds of the eye's refractive power [1]. The cornea is made up of three distinct layers: outer epithelium, stroma, and inner endothelium, which in turn are separated by the Bowman's and Descemet's membranes (Fig. 7.1).

The nonkeratinizing stratified corneal epithelium serves as the principal physical barrier to fluid and pathogens, which it does by forming an orderly arranged structure with strong intracellular tight junctions and by constant replacement of shed superficial epithelial cells by deeper, suprabasal epithelial cells. These are formed by the differentiation and migration of epithelial stem cells located in the basal region of the corneal limbus at the cornea–scleral junction [2]. The stroma comprises the majority (90%) of the corneal thickness and strength, consisting of regularly arranged arrays (lamellae) of collagen type I and V fibers, vital for corneal clarity, which interact with a number of different proteoglycans. The stroma is sparsely populated by embedded keratocytes which secrete this matrix [3]. The posterior cornea is composed of Descemet's membrane which acts as a specialized basement membrane for the monolayer of nondividing endothelial cells. These nondividing cells' function is essential for stromal dehydration through tight junctions and endothelial pumps [2]. Dysfunction of any of the elements that make up the cornea will result in a loss of corneal clarity and, ultimately, sight loss.

Corneal dystrophies

Corneal dystrophies are a group of inherited, heterogeneous, bilateral disorders that affect the transparency or shape of the cornea [4]. Traditionally, these dystrophies were classified according to the corneal layer most affected, although this classification did not adequately address those corneal dystrophies in which more than one layer is affected or where the layer affected changes as the disease progresses. Advances in genetic analysis, which led, in many cases, to the identification

CRISPR Genome Surgery in Stem Cells and Disease Tissues
DOI: https://doi.org/10.1016/B978-0-12-817876-8.00006-1

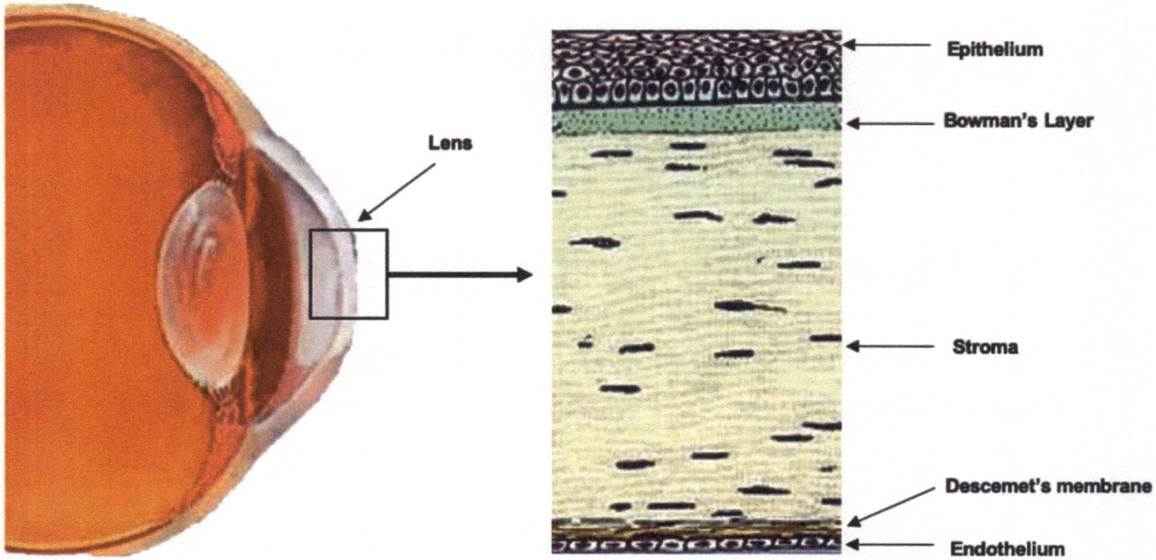

FIGURE 7.1 **The cornea.** The cornea is an avascular transparent tissue found in the anterior segment of the eye. It consists of five layers including the epithelium, Bowman's layer, stroma, Descemet's membrane, and endothelium. Due to its accessibility, small surface area and immune-privileged status the cornea is an ideal tissue for genome editing strategies.

of the causative genes, have deepened our understanding of the molecular mechanisms of this disease and revealed the extensive genotypic and phenotypic heterogeneity that exists. In 2008, and again in 2015, the integration of information on genotype, phenotype, and pathology enabled a reclassification of corneal dystrophies by the International Committee for Classification of Corneal Dystrophies that aimed to preserve the traditional grouping while making way for the new era of genetic advancements [5]. Dystrophies are now divided into epithelial and subepithelial dystrophies, epithelial–stromal transforming growth factor beta-induced (TGFBI) dystrophies, stromal dystrophies, and endothelial dystrophies (Table 7.1).

The age of onset and severity of the dystrophic phenotype can vary substantially, and the treatment strategy selected is tailored to the individual patient [4]. Corneal dystrophy may remain, without a precipitating trauma or injury [6,7], asymptomatic, or in other instances opacities appearing from early childhood reduce visual acuity and ultimately result in complete loss of vision. The most effective treatment choice for the patient is made depending on the current stage of the dystrophy [8]. For milder cases, conservative therapies such as gels/ointments, therapeutic contact lenses, and/or mild corneal abrasion are suitable. More severe cases require surgical interventions that are based on the location of the opacities within the cornea. Phototherapeutic keratectomy can be considered for superficial epithelial dystrophies and, with less success, stromal dystrophies. However, this approach, treating the symptoms and not the cause, is not curative and often offers only a temporary solution, likely requiring repeated treatments [9–12]. Unfortunately, in many instances sight deteriorates to the point where a corneal transplantation is required and while well-established, drawbacks including the shortage of corneal donors and graft rejection following transplant

TABLE 7.1 Corneal dystrophies: associated inheritance pattern, gene locus, and causative genes.

Category 1: A well-defined corneal dystrophy in which the gene has been mapped and identified and the specific mutations are known.
Category 2: A well-defined corneal dystrophy that has been mapped to one or more specific chromosomal loci, but the gene(s) remains to be identified.
Category 3: A well-defined corneal dystrophy in which the disorder has not yet been mapped to a chromosomal locus.
Category 4: This category is reserved for a suspected, new, or previously documented corneal dystrophy, although the evidence for it, being a distinct entity, is not yet convincing.

From J.S. Weiss, H.U. Møller, A.J. Aldave, B. Seitz, C. Bredrup, T. Kivelä, et al., IC3D classification of corneal dystrophies—edition 2, Cornea 34 (2) (2015) 117–159 [5].

limit its use [13]. However, in the case of some corneal dystrophies, such as those caused by mutations in the *TGFBI* gene, a reemergence, or in some cases a novel occurrence, of mutant protein in the corneal graft occurs [14–18]. An approach that tackles the underlying genetic cause of corneal dystrophy with a minimally invasive technique would present a very attractive option [19].

Gene therapy in the cornea

The cornea offers the ideal candidate for targeted gene therapy due to its small surface area, accessibility, and ease of visualization [19]. In addition, the unique immune-privileged status of the cornea [20] minimizes immune response to the gene-based therapeutics and delivery vehicles that are introduced. Furthermore, the avascularity of the cornea minimizes the risk that any of the administered gene therapy and delivery agents will escape the cornea to reach extraocular sites and cause off-target events in other tissues.

Conventionally, gene therapy refers to the introduction of a functional copy of the gene, usually as a viral transgene, to treat loss-of-function mutations (Fig. 7.2A).

This allows the target cell to produce a functional protein in cases when the endogenous protein is defective. The use of such an approach has been successfully applied to treat recessive, loss-of-function, heritable diseases of the retina [21]. For example, Leber's congenital amaurosis type II (LCA2), due to loss-of-function mutations in the *RPE65* gene, has been treated by a subretinal injection of *RPE65* cDNA packaged in adeno-associated virus-2 (AAV-2) [21] and was the first reagent to receive FDA approval [22].

Such an approach for the treatment of dominant disorders is more problematic since expression of the mutant gene product persists alongside the expression of the functional transgene (Fig. 7.2B). Whether this approach works may depend upon the mechanism of action of the dominant mutation. For instance, milder forms of retinitis pigmentosa (RP) may be caused by dominant-negative mutations in which the mutant protein forms a dimer with normal rhodopsin and interferes with its function or assembly in the disk membrane. In contrast, more severe forms of RP may be caused by gain-of-function mutations, toxic to the rod cell either through mislocalization or by continuous activation of the unfolded protein response [23]. The balance between functional homodimers and nonfunctional dominant-negative rhodopsin heterodimers may be tipped by increasing the level of normal protein through transgene expression [23]. However,

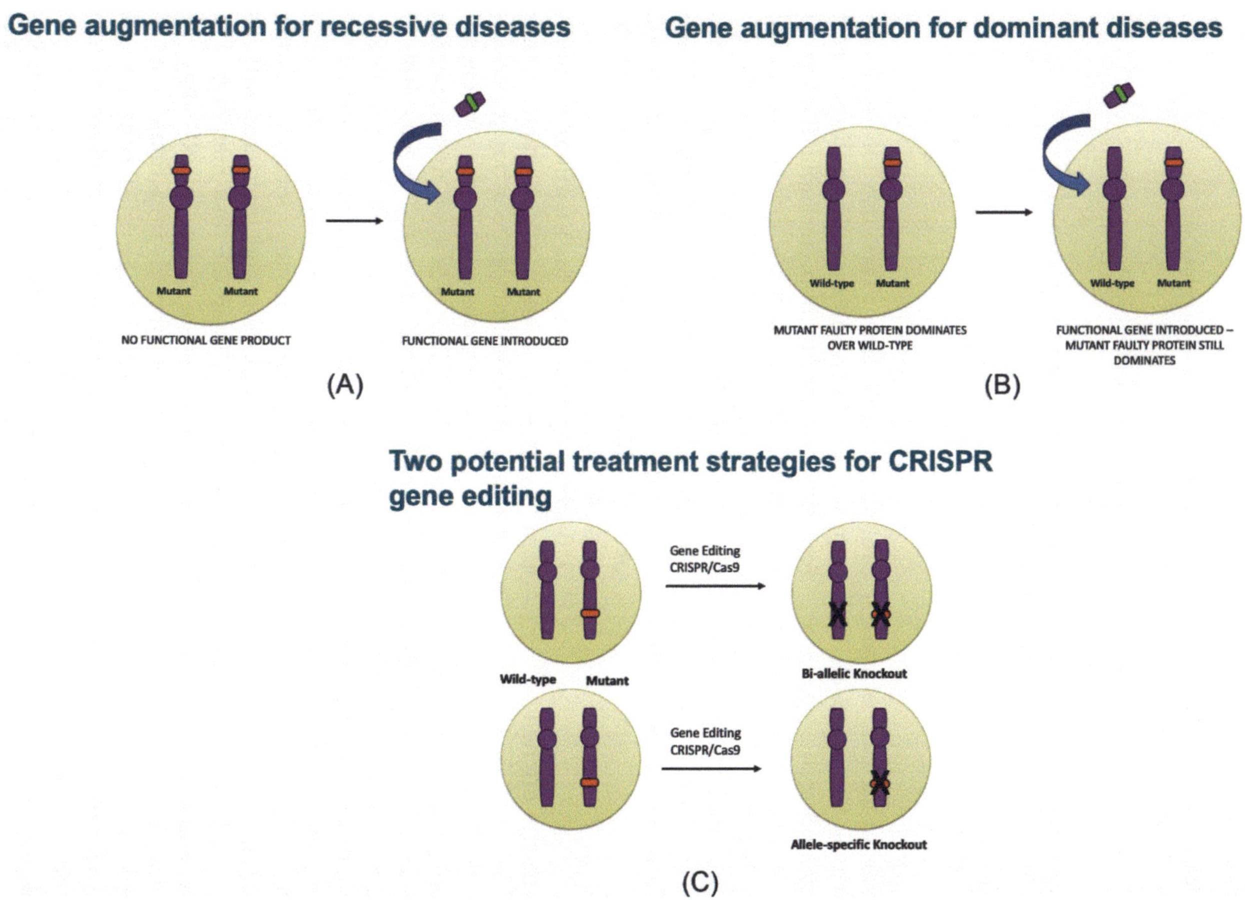

FIGURE 7.2 **Strategies for gene therapy in autosomal recessive and dominant disease.** (A) In recessive disorders, introduction of a transgene encoding wild-type protein allows the target cell to produce a functional protein in cases when the endogenous protein is absent or defective. (B) In dominant disorders, expression of the mutant gene product persists alongside the expression of the functional protein by the introduced transgene. (C) Biallelic gene disruption abolishes expression from both the wild-type and mutant alleles. A guide sequence can be chosen that can be used for all known mutations using criteria of high on-target and very low off-target cutting in which the guide sequence does not overlap with any of the known mutations. Mutation-dependent allele-specific genome editing can be achieved by selecting guides that place the mutation either in the protospacer adjacent motif (PAM) site (PAM-specific) or in the 20 bp single guide RNA (sgRNA) guide sequence (guide-specific). *CRISPR*, clustered regularly interspaced short palindromic repeat.

the expression of injurious mutant gain-of-function mutations may need to be suppressed in order to achieve therapeutic benefit from the virally delivered transgene. Suppression of expression may be achieved in an allele-specific manner [24] or by blocking expression of both endogenous wild-type and mutant alleles and delivery of a suppression-resistant replacement transgene [25,26].

Meesmann epithelial corneal dystrophy (MECD) has been shown to result from cellular toxicity caused by mutant keratin 12 protein induction of the unfolded protein response [27], while mutant TGFBI protein has been shown to accumulate in the corneal deposits seen in granular and lattice TGFBI corneal dystrophies (TGFBICD) [28]. This suggests that, in both cases, suppression of expression of the mutant protein is required.

The autosomal dominant inheritance pattern of the *TGFBI* corneal dystrophies suggests that selective ablation of only the mutant allele may be a viable treatment strategy [19,29]. Heterozygous nonsense mutations in *TGFBI* associated with a normal phenotype have been reported [30,31] indicating that *TGFBI* is haplosufficient and disruption of the mutant allele alone would not lead to a more severe phenotype. Both heterozygous null *TGFBI* and *Krt12* mice have been reported to be phenotypically normal [32,33]. However, while the homozygous null *TGFBI* mouse has been shown to display only small changes in corneal structure [34], wound healing in a zebrafish tail fin regeneration assay was inhibited following *tgfbi* morpholino injection (unpublished observations). This observation indicates an essential role of TGFBIp in wound repair and suggests that maintenance of wild-type TGFBIp in the cornea is required to maintain normal repair responses. Biallelic knockout of *TGFBI* could be detrimental to corneal healing, during which TGFBIp expression is known to be upregulated, and thus allele-specific gene disruption is necessary, in which only the mutant allele is targeted for disruption, leaving the wild-type allele intact.

The potential of therapeutic small interfering RNA (siRNA) therapy to suppress mutant protein expression for autosomal dominant corneal dystrophies has been clearly demonstrated both for MECD and TGFBICD [35–39]. Alternatively, biallelic knockdown of *TGFBI* expression has been demonstrated in human corneal cell lines [40,41]. However, siRNA-induced knockdown of mutant protein expression, limited by the problematic uptake of therapeutic levels of these charged molecules by the intact cornea, would require daily application of siRNA [42]. Furthermore, a different siRNA would be required for each of the 70 *TGFBI*, 23 *KRT12*, and 3 *KRT3* known mutations.

CRISPR/Cas9 gene editing for corneal dystrophies

Clustered regularly interspaced short palindromic repeat (CRISPR)/Cas9 gene editing provides a gene therapy approach that could provide a permanent or longer lasting suppression of mutant protein expression. Cas9 can be directed to cut a desired sequence in the genome—provided it is directly upstream of a protospacer adjacent motif (PAM)—by simply altering the guide RNA (sgRNA) sequence. The site-specific sgRNA will direct the Cas9 nuclease to make a double-strand break (DSB) and the cell will then attempt to repair this damage either by precise homology-directed repair (HDR) or by error-prone nonhomologous end joining (NHEJ) [43]. By harnessing these cellular responses, different forms of gene editing can be achieved.

Homology-directed repair

In HDR, the DSB is repaired using an exogenous homologous repair template, added together in large excess with the CRISPR/Cas9 gene editing components, to encourage repair mechanisms to introduce desired sequence changes and achieve gene correction. However, as homologous recombination is restricted to the S and G2 phases of the cell cycle, gene correction via HDR is limited to dividing cells [44]. This approach has been successfully demonstrated in ex vivo repair of the *TGFBI* gene in human corneal keratocytes derived from a granular corneal dystrophy patient [45]. However, the efficiency of HDR is currently too low to be applicable for local in vivo administration.

Nonhomologous end joining repair

Two alternative approaches to the use of NHEJ-induced repair errors by CRISPR/Cas9 in the treatment of corneal dystrophies can be envisaged (Fig. 7.2C). In the first, biallelic gene disruption abolishes expression from both the wild-type and mutant alleles. A guide sequence can be chosen that can be used for all known mutations using criteria of high on-target and very low off-target cutting in which the guide sequence does not overlap with any of the known mutations. Such an approach has been used to generate *TGFBI* knockout limbal epithelial stem cells (LESCs) in which biallelic disruption of *TGFBI* has been demonstrated [46]. In practice, LESCs, obtained from a patient's limbal biopsy, would be genetically altered ex vivo, and subsequently reintroduced into the patient's cornea. However, in order for the clonal edited cells to repopulate the cornea, resident stem cells still bearing the mutation would have to be completely removed. Consequentially, this action would result in a state of limbal stem cell deficiency, a severe condition that leads to corneal conjunctivalization, neovascularization, epithelium instability, and visual loss [47]. The risk that the graft of clonally edited LESCs would fail may be considered too high, thus this would not be a feasible treatment strategy for the corneal dystrophies. However, an approach that selectively disrupts the mutant allele and leaves the wild-type allele intact would be a viable treatment option.

Mutation-dependent allele-specific genome editing

Mutation-dependent allele-specific genome editing can be achieved by harnessing the mechanism of action of the RNA-guided Cas9 nuclease. The Cas9 nuclease can be redirected to a desired target sequence simply by altering the 20-nucleotide spacer sequence within the sgRNA, provided it lies directly upstream of a PAM, recognized directly by the Cas9 protein [48,49]. PAM recognition is the initial step in Cas9 binding to a target site, the PAM interacting domain samples the PAM. We exploited this dependence upon PAM recognition to demonstrate, for the first time, allele-specificity of gene editing where the disease-causing missense mutation generated a PAM recognition motif not present in the wild-type sequence [50]. Thus, an MECD-causing mutation in *Krt12* results from a single base pair T > C transition mutation. This alters the wild-type sequence from CTT (Leu) to the mutant CCT (Pro), and on the opposite strand, this alters the sequence from a weak noncanonical PAM (AAG) to a strong Cas9 PAM (AGG). A sgRNA targeting Cas9 to this site was found to be highly effective at reducing the expression of mutant *KRT12* mRNA and protein in vitro. Subsequently, when a combined Cas9/sgRNA expression construct was injected into the corneal stroma of a humanized MECD mouse model [27], resulted in NHEJ repair-induced frame-shifting deletions specifically within the mutant *KRT12* allele [50] (Fig. 7.3A).

This approach was further validated in a lattice corneal dystrophy-associated *TGFBI*

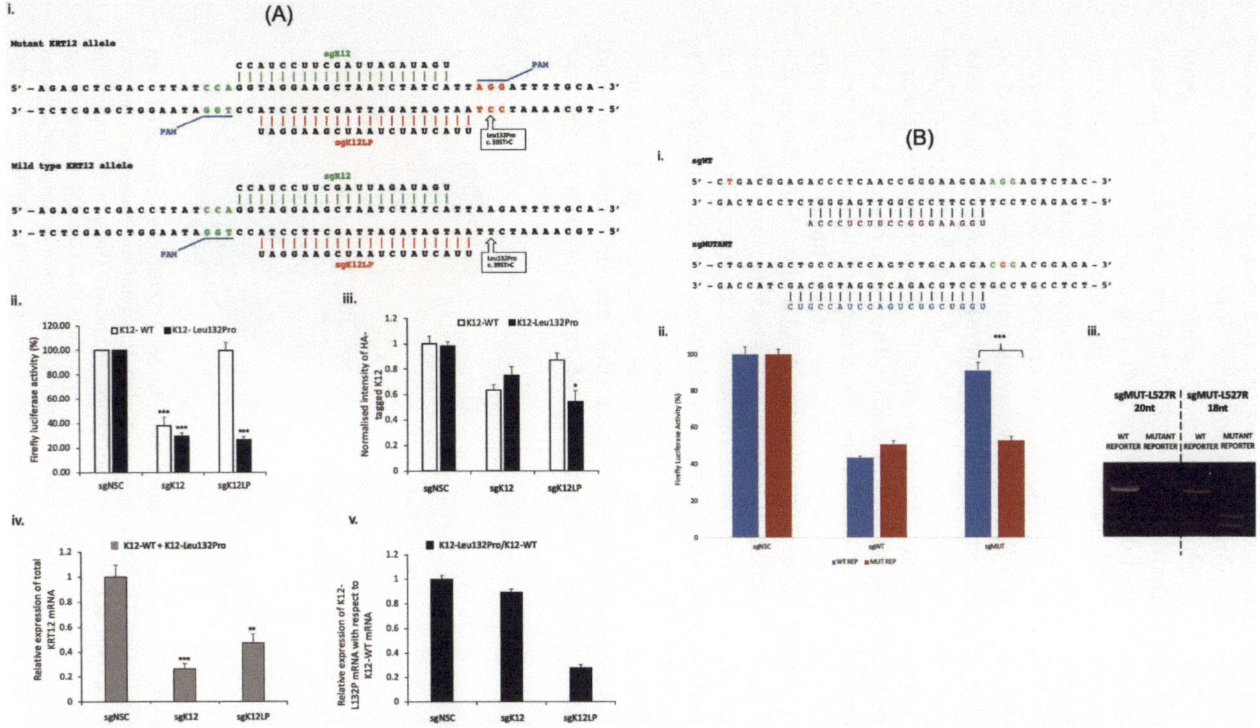

FIGURE 7.3 **Allele-specific cleavage of mutations utilizing a protospacer adjacent motif (PAM)-specific approach.** (A) Evaluation of allele specificity and potency of sgK12LP using exogenous constructs. Exogenous expression constructs for wild-type and mutant K12 were used to test the allele-specificity and potency of sgK12LP. (i) Design of single guide RNAs (sgRNAs) for targeting wild-type and mutant *K12* alleles. A sgRNA to use the SNP-derived PAM found on the *K12–L132P* allele was designed (*red*). This PAM is absent in the wild-type allele. A second sgRNA to target both wild-type and mutant *K12* alleles (*green*) was also designed and used as a positive control. (ii) A dual-luciferase assay demonstrated the allele specificity of the sgK12LP plasmid, whereas potency was shown to be comparable to that of the sgK12 construct. $N = 8$. (iii) Western blotting further demonstrated these attributes with a noticeable reduction in K12–L132P protein in cells treated with sgK12LP in comparison with cells treated but expressing K12 wild-type protein. β-Actin was used as a loading control. (iv) Quantitative reverse transcriptase polymerase chain reaction for total K12 in cells expressing both wild-type and mutant alleles demonstrated a knockdown in mRNA expression. $N = 4$. (v) Allele proportions of this mRNA knockdown were then quantified by pyrosequencing, confirming an allele knockdown of the mutant allele in cells coexpressing both *KRT12* alleles and treated with sgK12LP. $N = 4$, $^{*}P < .05$, $^{**}P < .01$, $^{***}P < .001$. (B) (i) The L527R mutation (c.1580 T > G) is indicated in red and PAM utilized is shown in green. A 20-nucleotide (nt) sgRNA targeted to a naturally occurring PAM was designed as a positive control (sgWT, *purple*—top of figure). A 20-nt sgRNA utilizing the novel PAM, containing the L527R mutation, was designed sgMUTANT, *blue*—bottom of figure). (ii) Both sgWT and sgMUTANT were targeted to a luciferase reporter plasmid containing either a wild-type or mutant transforming growth factor beta-induced (*TGFBI*) gene sequence to determine potency and allele specificity. (iii) An in vitro digestion with Cas9 protein complexed with a sgRNA utilizing the novel L527R PAM was carried out to confirm the specificity observed. Mutant guides of both 20 and 18 nts were tested. *Reproduced with permission from (A) D.G. Courtney, J.E. Moore, S.D. Atkinson, E. Maurizi, E.H.A. Allen, D.M.L. Pedrioli, et al., CRISPR/Cas9 DNA cleavage at SNP-derived PAM enables both in vitro and in vivo KRT12 mutation-specific targeting, Gene Ther. 23 (2016) 108–112 (Figures 1 and 2) [50]. (B): K.A. Christie, D.G. Courtney, L.A. DeDionisio, C. Chao-Shern, S. De Majumdar, L.C. Mairs, et al., Towards personalised allele-specific CRISPR gene editing to treat autosomal dominant disorders, Sci. Rep. 7 (2017) 16174 (Figure 4) [29].*

mutation (c.1580 T > G, L527R) [51] that generates a novel PAM (CTG > CGG). Specificity was confirmed by both a previously described in vitro dual-luciferase assay [35,37] and by an in vitro DNA cleavage assay [19,29] (Fig. 7.3B).

An alternative to the PAM-specific approach is offered by the next stage in target recognition by the Cas9-sgRNA ribonucleoprotein. Once the correct PAM has been identified the region of the guide proximal to the PAM will direct target binding [52–54]. The PAM proximal region, consisting of the first 5–12 bp of the guide sequence [54–56] has been coined the "seed region." As the PAM proximal region dictates whether Cas9 binds to the target site, mismatches in this region are critical in determining specificity of Cas9. If the PAM proximal region is complementary to the guide sequence, the PAM distal region will then bind. Although complementarity between the PAM distal region and guide sequence is required, reports have shown mismatches here are better tolerated than in the PAM proximal region [54–56]. Thus should the disease-causing missense mutation be located within this PAM proximal region it may be possible to distinguish between the wild-type and mutant alleles and disrupt the mutant allele while leaving the wild-type allele intact.

More than 60 missense mutations have been identified in *TGFBI* that cause a variety of corneal dystrophies [14] (Fig. 7.4A).

These mutations span the entire *TGFBI* gene but are clustered in hotspots found in exon 4, 11, 12, and 14. An analysis of these 60 known *TGFBI* missense mutations was performed to determine if they generate a novel *Streptococcus pyogenes* Cas9 PAM or if a *S. pyogenes* Cas9 PAM is found within the 8 nucleotides, placing the mutation within the seed region [29]. Out of 60 mutations, 19 generate a novel *S. pyogenes* PAM, while 44 out of 60 have a naturally occurring adjacent PAM site that places the mutation within the seed region. When these figures are considered together, 20% of the *TGFBI* missense mutations are not targetable by *S. pyogenes* Cas9 (Fig. 7.4B).

Despite the wide spectrum of mutations, most are found in only one or a few cases and the vast majority of cases are due to the five

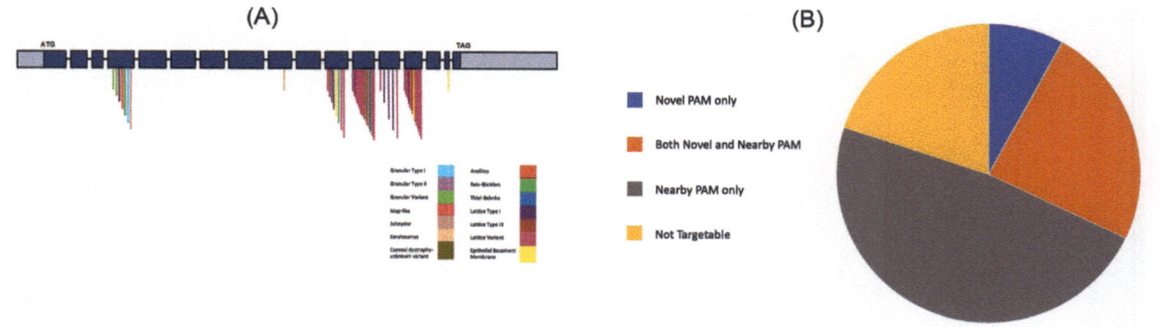

FIGURE 7.4 **Over 60 different mutations in transforming growth factor beta-induced (*TGFBI*) gene associated with corneal dystrophies.** (A) Schematic of the 17 exon *TGFBI* gene. Coding exons are shaded dark blue and noncoding regions are shaded light blue. The location of the mutations is indicated below and color-coded, as shown. Hotspots for mutations are clearly seen in exons 4, 11, 12, and 14. (B) Analysis of TGFBI corneal dystrophy mutations in a CRISPR system. Venn diagram to illustrate the total number of TGFBI mutations that generate a novel *Streptococcus pyogenes* protospacer adjacent motif (PAM), have a near-by *S. pyogenes* PAM, that is, within the first 8 bp of the guide sequence, have both a novel and near-by *S. pyogenes* PAM or are not targetable by either approach. *Reproduced with permission from K.A. Christie, D.G. Courtney, L.A. DeDionisio, C. Chao-Shern, S. De Majumdar, L.C. Mairs, et al., Towards personalised allele-specific CRISPR gene editing to treat autosomal dominant disorders, Sci. Rep. 7 (2017) 16174 (Figure 3) [29].*

most prevalent mutations found in either codon 124 (exon 4) (R124C, R124H, R124L) or codon 555 (exon 12) (R555Q and R555W) (Table 7.2).

Remarkably, each of these mutations, differing by only a single amino acid, results in strikingly different protein aggregates with a very strong genotype–phenotype correlation. While none of these five mutations generate a novel PAM, the missense mutation is present in the seed region of the guide sequence in all cases. To achieve complete allele-specificity for a particular *TGFBI* mutation, stringent fidelity is required as an almost perfect off-target site exists in the form of the wild-type allele, which differs by only one base pair from the mutant. Therefore, we investigated if a guide-specific approach could confer stringent allele-specificity [29]. We found that *S. pyogenes* Cas9 was not able to adequately discriminate between wild-type and mutant alleles differing in sequence at only one base of the guide sequence (Fig. 7.5).

Reports which indicated that truncating the guide RNA can improve specificity, led us to perform a screen of guide length from 16 to 22 bp (Fig. 7.6).

However, no significant improvement in specificity was observed. Interestingly, we found that mismatches closer to the PAM conferred more discrimination between alleles than PAM distal mismatches, confirming previous reports [54–56]. However, even though PAM-proximal mismatches result in superior discrimination, they do not confer enough specificity to be suitable for gene therapy.

Mutation-independent allele-specific genome editing

The drawbacks of a mutation-dependent gene therapy approach to corneal dystrophies described above include the large number of different *TGFBI* mutations, each of which would require a separate, stringently verified guide; less than one-third of *TGFBI* mutations are targetable by a PAM-specific approach; lack of mutant allele specificity achieved by a guide-specific approach; and one-fifth of *TGFBI* mutations are targetable by neither approach. Taken together, these limitations have led us to investigate the utility of a mutation-independent allele-specific approach. We have already demonstrated that excellent allele-specificity can be achieved between wild-type and mutant alleles differing only by a single base pair when a missense mutation generates a novel PAM site. Similarly, single nucleotide polymorphisms (SNPs) can be associated with a PAM on one allele and not on the other [57,58]. This occurrence could be used as a proxy for disease-causing mutations if they lie in proximity to the mutation on the same chromosome (in cis). Lee et al. reported the use of the extensive knowledge of the Huntington (*HTT*) gene haplotypes [59], encompassing the trinucleotide repeat associated

TABLE 7.2 The five most prevalent *TGFBI* mutations occur in two codons and cause five phenotypically distinct corneal dystrophies.

Mutation	Associated corneal dystrophy	Codon change
R124C	Classic lattice type 1	CGC > TGC
R124H	Granular type 2 (Avellino)	CGC > CAC
R124L	Reis–Bucklers	CGC > CTC
R555Q	Thiel–Behnke	CGG > CAG
R555W	Classic granular type 1	CGG > TGG

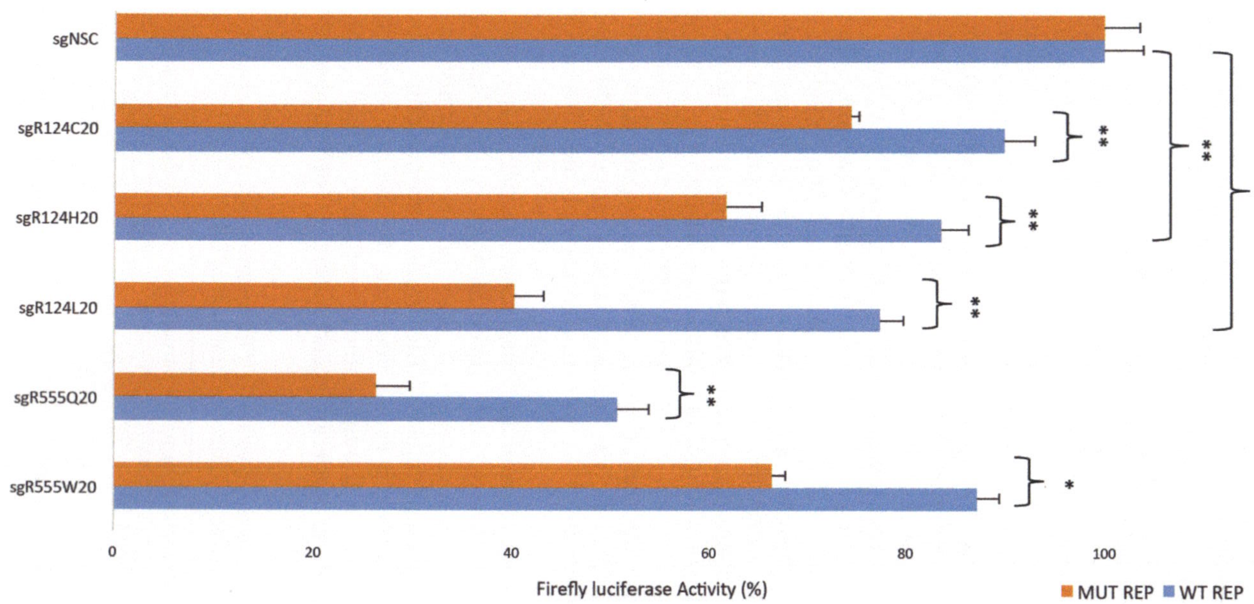

FIGURE 7.5 **Investigation of a guide-specific approach to treat prevalent transforming growth factor beta-induced (*TGFBI*) gene mutations.** Using a guide-specific approach, 20-nucleotide guides for the five most prevalent *TGFBI* mutations were targeted to wild-type and respective mutant sequence in a dual-luciferase assay. The five guides cut with varying degrees of specificities and efficiencies. There was a significant difference between the wild-type and mutant sequence in all cases. *Source: Adapted from K.A. Christie, D.G. Courtney, L.A. DeDionisio, C. Chao-Shern, S. De Majumdar, L.C. Mairs, et al., Towards personalised allele-specific CRISPR gene editing to treat autosomal dominant disorders, Sci. Rep. 7 (2017) 16174 (Figure 6) [29].*

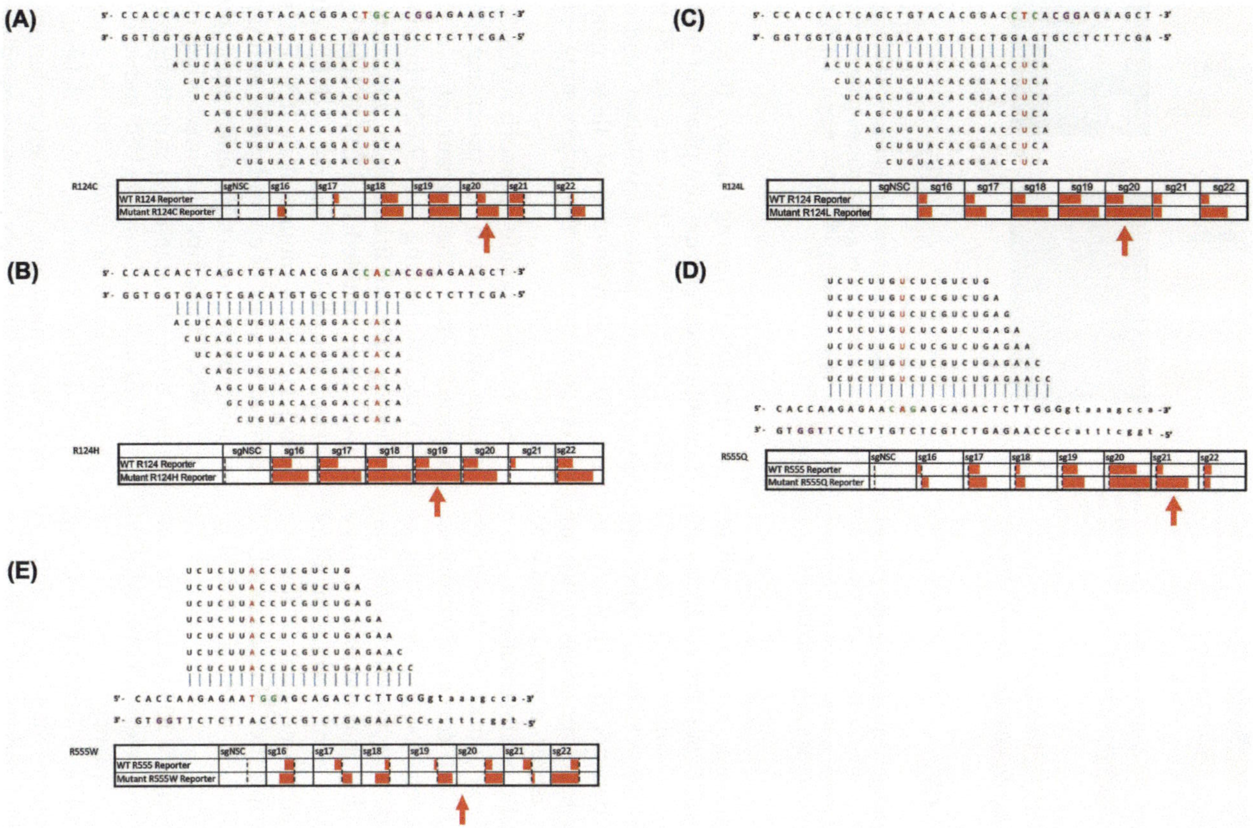

FIGURE 7.6 Guide-length screen to determine the effect on specificity of a guide-specific system. Heatmaps showing varying degrees of knockdown observed via a dual-luciferase assay when guides ranging in different lengths are targeted to the wild-type and respective mutant sequence. Specificity bars show knockdown when normalized to the nonspecific control, with 100% being maximal knockdown observed. Maximal allele specificity observed for each mutation indicated with a red arrow. (A) R124C, (B) R124H, (C) R124L, (D) R555Q, (E) R555W. *Source: Adapted from K.A. Christie, D.G. Courtney, L.A. DeDionisio, C. Chao-Shern, S. De Majumdar, L.C. Mairs, et al., Towards personalised allele-specific CRISPR gene editing to treat autosomal dominant disorders, Sci. Rep. 7 (2017) 16174 (Figure 7) [29].*

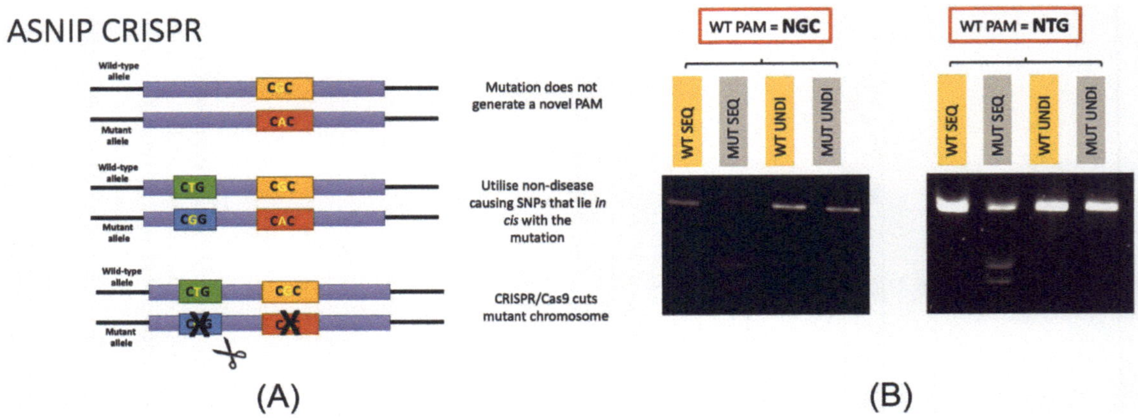

FIGURE 7.7 Mutation-independent allele-specific gene editing. (A) Schematic representation to illustrate the concept of allele-specific, SNP-derived, in cis, personalized (ASNIP) CRISPR. (B) Activity and specificity of guides targeting the in cis SNPs was verified by in vitro DNA cleavage assays. *CRISPR*, Clustered regularly interspaced short palindromic repeat; *PAM*, protospacer adjacent motif; *SNPs*, single nucleotide polymorphisms; *WT*, wild type.

with Huntington's disease, to achieve allele-specific disruption of *HTT* [60]. For *HTT*, screening of only 10 SNPs is sufficient to provide haplotype information for 97% of the European Huntington's disease patients [61]. However, in the case of *TGFBI* missense mutations which are spread over a much larger genomic distance, phase cannot be as easily predetermined.

Building on previous work, we developed an approach for the *TGFBI* corneal dystrophies that would enable allele-specific targeting of the *TGFBI* locus of a patient harboring any missense mutation, provided that the patient has a heterozygous haplotype across some or all of the *TGFBI* locus. The name coined for this approach was allele-specific, SNP-derived, in cis, personalized, CRISPR (ASNIP CRISPR) (Fig. 7.7A).

Around 24 SNPs across a 35 kbp region containing the *TGFBI* gene were identified that have a minor allele frequency of >0.1 in the 1000 Genomes Project Phase 3 cohort (ensuring a reasonable probability that an individual would be heterozygous) and contain a PAM on only one allele. Activity and specificity of guides targeting these SNPs was verified by in vitro DNA cleavage assays (Fig. 7.7B). It is possible to determine the phase of these PAM-associated SNPs relative to a *TGFBI* missense mutation in a patient by linked-read sequencing [61]. Pairwise use of these guides may direct deletion of the intervening region of DNA (Fig. 7.8).

Such a dual guide approach to CRISPR/Cas9 editing was shown to be a promising therapeutic approach for the treatment of patients with LCA10 bearing the *CEP290* intronic splice mutation [62]. Additionally, in a human cell line, it was demonstrated that DNA deletions up to 10 kbp, coupled to precise NHEJ, effectively knockeddown a mutant allele [63]. Excision of a defined number of exons by this technique gives a much more predictable effect upon the resulting mRNA transcript produced when compared to the unpredictable indel mutations introduced by NHEJ repair of a single DSB induced by a single guide. Pairs of guides can be chosen that, when the intervening DNA is deleted, cause a frameshift mutation, premature termination of translation, and likely nonsense-mediated decay of the mutant transcript [64,65]. Deletion of the intervening genomic region was demonstrated

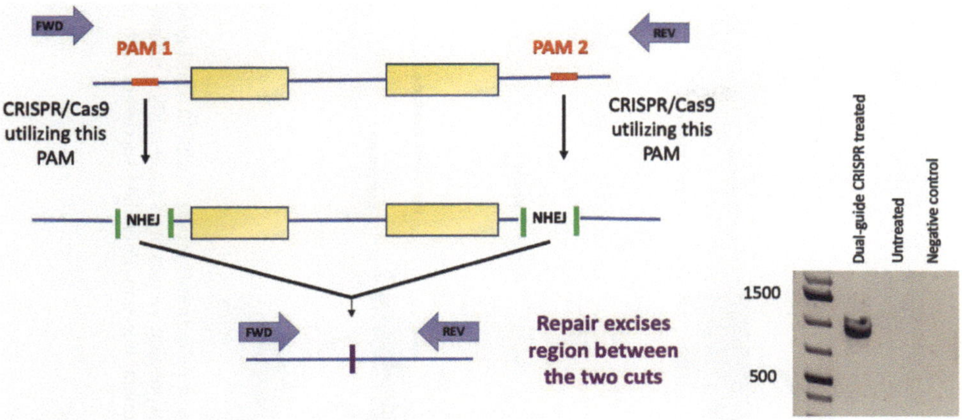

FIGURE 7.8 Schematic to show method of detection of deletion of a portion of the transforming growth factor beta-induced (*TGFBI*) gene following dual-cut clustered regularly interspaced short palindromic repeat (CRISPR)/Cas9 gene editing. Prior to, or in the absence of, deletion, forward and reverse primers are too far apart for polymerase chain reaction (PCR) amplification. Deletion of the intervening region, defined by the two protospacer adjacent motif (PAM) sites utilized, brings the primers closer together and this can be detected by PCR.

by cotransfection of a cell line derived from a patient harboring a R124H *TGFBI* mutation with Cas9-sgRNA RNPs for two of the selected PAM-associated SNPs (Fig. 7.7).

Delivery of CRISPR/Cas9 gene editing components to the cornea

Despite its accessibility, small size, and immune-privileged status, delivery of gene therapy components to the different layers of the cornea has proven to be challenging [66] (Fig. 7.9).

The main obstacles to achieving a therapeutic dose, by the diffusion of a drug through the eye surface, are the protective mechanisms and underlying ocular anatomy [67,68]. Topically applied drugs may be eliminated from the ocular surface by various mechanisms including lacrimation and tear turnover, drug metabolism, and preferential conjunctival absorption [69]. The next barrier to the entry of gene therapy components is the cornea epithelium in which the cells are linked by characteristically impermeable tight junctions [70]. In addition to the biological barriers that exist, the entry of large macromolecules, such as DNA or siRNA, into corneal cells is challenging. These large, negatively charged, hydrophilic molecules do not easily cross the nonpolar cell membrane [71]. Molecules can traverse the corneal epithelium along the transcellular or paracellular pathways, but this requires DNA or siRNA molecules to navigate the exceptionally tight junctions of the corneal epithelium [70]. Development of a noninvasive delivery method such as self-administrated eye drops has proven challenging [72,73]. Delivery agents need to overcome the tear barrier and remain associated with the corneal epithelium long enough to allow cellular internalization and subsequent release to function within the target cell with optimal bioavailability. While commercially available transfection agents, such as lipofectamine and polyethyleneimine, have proven unable to deliver siRNA to mouse cornea in vivo [72], we have recently demonstrated functional topical delivery of siRNA noncovalently complexed with a modified derivative of a cell penetrating peptide. This technique resulted in significant in vivo

	Plasmid	AAV	RNPs
Topical Application	No	No	No
Intrastromal Injection	Yes	Yes	Yes
Intracameral Injection	No	Yes	No

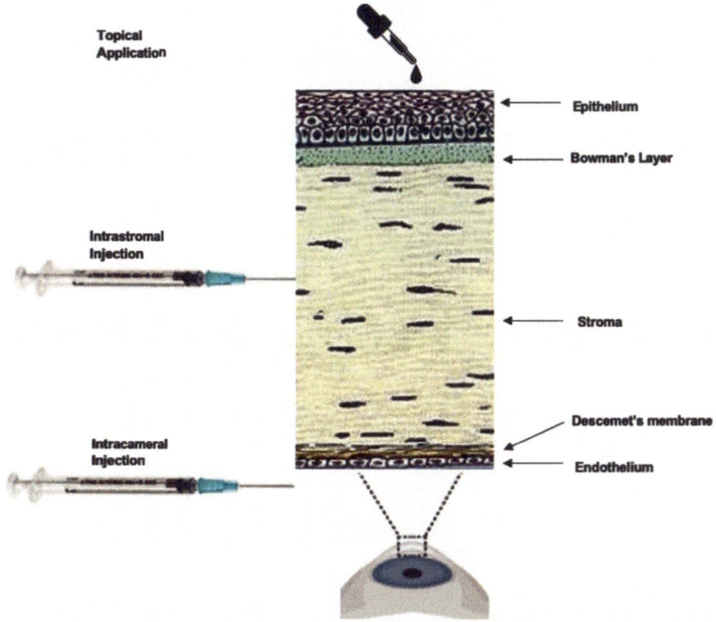

FIGURE 7.9 **Delivery of gene therapy components to the cornea.** Clustered regularly interspaced short palindromic repeat (CRISPR)/Cas9 gene therapy components can be delivered to the cornea as either plasmid or adeno-associated virus (AAV) expression vectors or as Cas9-sgRNA ribonucleoprotein complexes (RNPs). Delivery can be made by topical application or by intrastromal or intracameral injection. Published success of each of these modes and routes of administration is indicated in the inset table.

knockdown of luciferase reporter gene expression in mouse corneal epithelium [42]. However, we have not been able to demonstrate functional delivery of CRISPR-Cas9 gene editing components, either as Cas9-sgRNA RNP, expression plasmid, or as viral vector, to the cornea by topical delivery.

Local administration via intrastromal or intracameral injection (Fig. 7.9) of gene therapy agents circumvents the problem of biological barriers. Thus injection of a coexpression plasmid vector for *Spy* Cas9 and a sgRNA targeting firefly luciferase reporter gene into the corneal stroma of a reporter knock-in mouse (*Krt12* + /

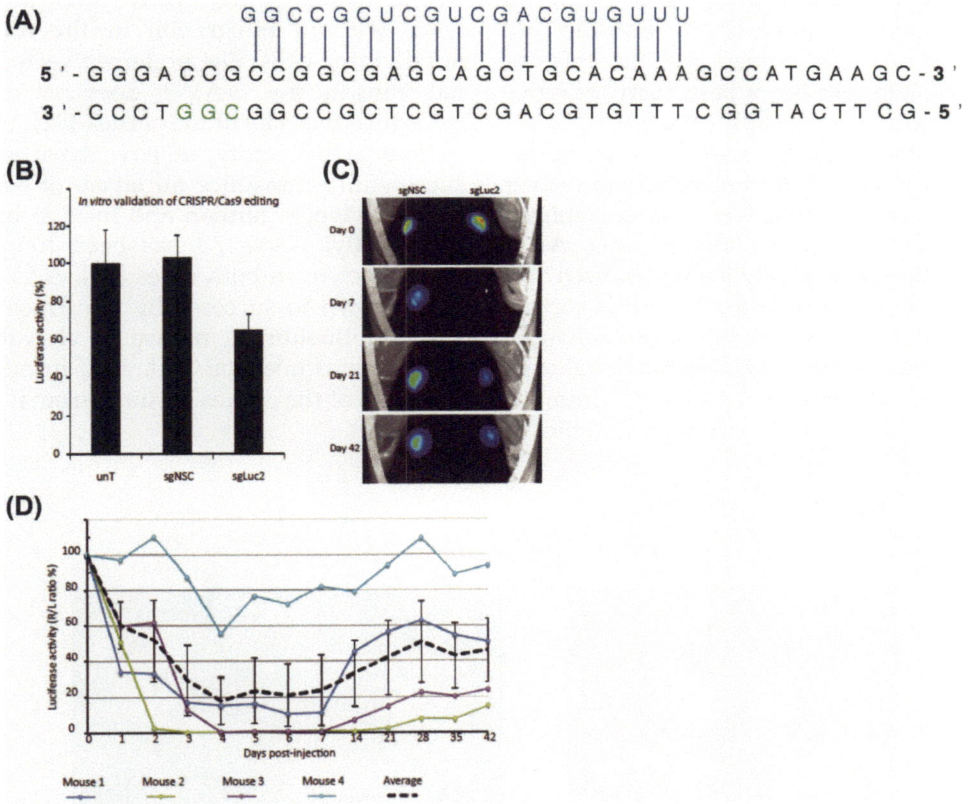

FIGURE 7.10 **Sustained clustered regularly interspaced short palindromic repeat (CRISPR)/Cas9 mediated silencing of luc2 in vivo.** (A) The short guide RNA (sgRNA) specific for luc2 was designed to target the 5′ region of the gene, to increase the likelihood of inducing a frameshifting deletion that would knock out luciferase activity by generating a premature termination codon. (B) An in vitro dual-luciferase assay demonstrated successful targeting of luc2 by the sgLuc2 construct, as shown by a significant reduction in luciferase activity when normalized to untreated cells (data normalized against the untreated control = 100%). (C) Representative image of mice displaying a maximal reduction in luc2 expression after injection with the sgLuc2 construct (right eye). This image was taken from the mouse represented by the green line in panel (D), below, at 7 days posttreatment. (D) After treatment, the corneal luciferase activity of each mouse was quantified using a Xenogen IVIS live animal imager every day for 7 days, then every 7 days thereafter, for a total of 6 weeks. Luciferase activity for each treatment group expressed as a percentage of control (R/L ratio %). *Source: Adapted from K.A. Christie, D.G. Courtney, L.A. DeDionisio, C. Chao-Shern, S. De Majumdar, L.C. Mairs, et al., Towards personalised allele-specific CRISPR gene editing to treat autosomal dominant disorders, Sci. Rep. 7 (2017) 16174 (Figure 2) [29].*

luc2)—that exclusively expresses firefly luciferase (luc2) in the corneal epithelium under control of the keratin K12 promoter—significantly reduces luciferase activity from the day after injection. Moreover, this injection resulted in maximal silencing of 82% ± 13%, partially sustained for more than 7 weeks (Fig. 7.10).

Although this technique is effective in the small mouse eye, where the pressure achieved during injection can force DNA into cells [74], when scaled up to a human eye, the entry of naked DNA into cells is not particularly effective due to the reduced achievable pressure.

The need for pressure transfection can be circumvented by the use of viral vectors to deliver gene editing components. Viral vectors utilized for gene therapy include adenoviruses, AAV, and retroviruses, specifically lentivirus. Each vector has been extensively tested for use in corneal transduction [75,76]. No single vector offers all desirable criteria: adenovirus boasts a large cargo capacity but initiates a substantial immune response [77,78]; AAV provides efficient transduction with a minimal immune response but has a limiting packaging capability [79,80]; and lentivirus has a considerable packaging capacity but can cause unwanted integration [81].

AAV has been the most widely used vector in ocular gene replacement therapy. The success of AAV in gene augmentation is largely due to its lack of immune response and persistence without integration in the nucleus of nondividing cells. The preferred vector for retinal transduction, AAV-2, does not efficiently transduce the cells of the cornea [82]. However, various AAV serotypes have been shown to successfully transduce all layers of the cornea both ex vivo in human and in vivo in mouse. Specifically, AAV-2/8 has been found to be most efficient in both cases [83]. AAV-2/9 has been shown to successfully transduce the cornea endothelium in mouse in vivo following intracameral injection [84], and, in our hands, all layers of the cornea by intrastromal injection (Fig. 7.11).

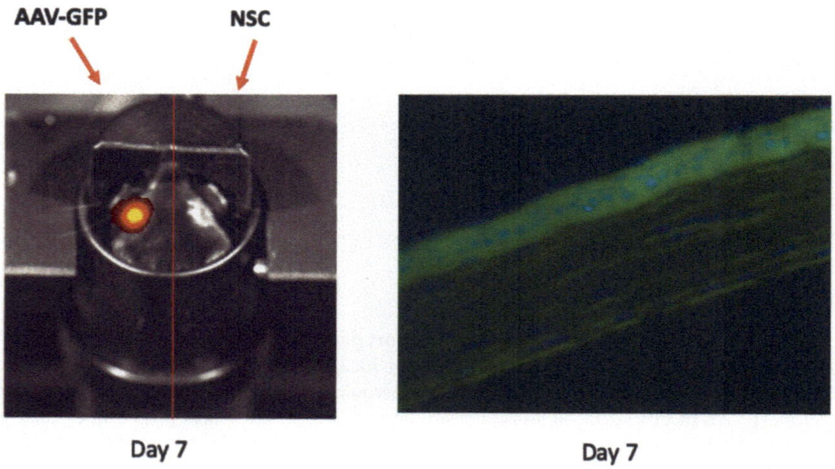

FIGURE 7.11 **Delivery of clustered regularly interspaced short palindromic repeat (CRISPR)/Cas9 components to the cornea by adeno-associated virus (AAV) vector.** The left eye of a mouse was injected intrastromally with AAV-2/9 containing either a green fluorescent protein (GFP) transgene or an empty vector. Around 7 days after injection, the corneal fluorescence was quantified using a Xenogen IVIS live animal imager. Eyes were collected, fixed in 4% paraformaldehyde in PBS, embedded in paraffin wax, sectioned, and visualized by fluorescence microscopy. GFP fluorescence was observed in all corneal layers from the epithelium (top) to endothelium (bottom).

Furthermore, systemic injection of AAV-2/9 results in transduction of the corneal epithelium, stroma, and endothelium [85]. Moreover, this injection appeared to transduce LESCs, as evidenced by sustained centripetal streaks of corneal epithelial cells emanating from the limbus.

Summary

While considerable progression toward a personalized genome editing strategy to treat the corneal dystrophies has been made, substantial work still remains to be done to overcome the current limitations of CRISPR/Cas9 gene editing, including limitations in the choice of PAM site, lack of global specificity, and inadequate delivery vehicles. These issues are not isolated to corneal gene therapy but extend to all gene therapy applications. Addressing these outstanding hurdles will enable the safe translation of these CRISPR/Cas9 gene editing therapies to the clinic.

References

[1] K.M. Meek, C. Knupp, Corneal structure and transparency, Prog. Retin. Eye Res. 49 (2015) 1–16.

[2] A.O. Eghrari, S.A. Riazuddin, J.D. Gottsch, Overview of the cornea: structure, function, and development, Prog. Mol. Biol. Transl. Sci. 134 (2015) 7–23.

[3] J.R. Hassell, D.E. Birk, The molecular basis of corneal transparency, Exp. Eye Res. 91 (2010) 326–335.

[4] G.K. Klintworth, Corneal dystrophies, Orphanet J. Rare. Dis. 4 (2009) 7.

[5] J.S. Weiss, H.U. Møller, A.J. Aldave, B. Seitz, C. Bredrup, T. Kivelä, et al., IC3D classification of corneal dystrophies – edition 2, Cornea 34(10) (2015) 117–159.

[6] C. Chao-Shern, R. Me, L.A. DeDionisio, B.L. Ke, M.A. Nesbit, J. Marshall, et al., Post-LASIK exacerbation of granular corneal dystrophy type 2 in members of a Chinese family, Eye (Lond.) 32 (2018) 39–43.

[7] H. Nithianandan, C. Chao-Shern, L. DeDionisio, T. Moore, C.C. Chan, Trauma-induced exacerbation of epithelial-stromal TGFBI lattice corneal dystrophy, Can. J. Ophthalmol. 54 (2019) e47–e49.

[8] B. Seitz, W. Lisch, Stage-related therapy of corneal dystrophies, Dev. Ophthalmol. 48 (2011) 116–153.

[9] M. Chen, L. Xie, Features of recurrence after excimer laser phototherapeutic keratectomy for anterior corneal pathologies in North China, Ophthalmology 120 (2013) 1179–1185.

[10] R. Dinh, C.J. Rapuan, E.J. Cohe, R. Laibson, Recurrence of corneal dystrophy after excimer laser phototherapeutic keratectomy, Ophthalmology 106 (1999) 1490–1497.

[11] A. Hafner, A. Langenbucher, B. Seitz, Long-term results of phototherapeutic keratectomy with 193-nm excimer laser for macular corneal dystrophy, Am. J. Ophthalmol. 140 (2005) 1–6.

[12] V.M. NoRathi, M. Taneja, S.I. Murthy, B. Bagga, P.K. Vaddavalli, V.S. Sangwan, Phototherapeutic keratectomy for recurrent granular dystrophy in postpenetrating keratoplasty eyes, Indian J. Ophthalmol. 64 (2016) 140–144.

[13] M. Bidaut-Garnier, E. Monnet, A. Prongué, R. Montard, A.S. Gauthier, M. Desmarets, et al., Evolution of corneal graft survival over a 30-year period and comparison of surgical techniques: a cohort study, Am. J. Ophthalmol. 163 (2016) 59–69.

[14] K.E. Han, S.I. Choi, T.I. Kim, Y.S. Maeng, R.D. Stulting, Y.W. Ji, et al., Pathogenesis and treatments of TGFBI corneal dystrophies, Prog. Retin. Eye Res 50 (2016) 67–88.

[15] A.J. Aldave, B. Sonmez, S.L. Forstot, S.A. Rayner, V.S. Yellore, B.J. Glasgow, A clinical and histopathologic examination of accelerated TGFBIp deposition after LASIK in combined granular-lattice corneal dystrophy, Am. J. Ophthalmol. 143 (2007) 416–419.

[16] R.M. Jun, H. Tchah, T.I. Kim, R.D. Stulting, S.E. Jung, K.Y. Seo, et al., Avellino corneal dystrophy after LASIK, Ophthalmology 111 (2004) 463–468.

[17] I. Jun, J.W. Jung, Y.J. Choi, T.I. Kim, K.Y. Seo, E.K. Kim, Long-term clinical outcomes of phototherapeutic keratectomy in corneas with granular corneal dystrophy type 2 excacerbated after LASIK, J. Refract. Surg. 34 (2018) 132–139.

[18] T.-I. Kim, M.I. Roh, H.E. Grossniklaus, S.J. Kang, S.M. Hamilton, D.F. Schorderet, et al., Deposits of transforming growth factor-beta-induced protein in granular corneal dystrophy type II after LASIK, Cornea 27 (2008) 28–32.

[19] C.B.T. Moore, K.A. Christie, J. Marshall, M.A. Nesbit, Personalised genome editing - the future for corneal dystrophies, Prog. Retin. Eye Res. 65 (2018) 147–165.

[20] J. Hori, T. Yamaguchi, H. Keino, P. Hamrah, K. Maruyama, Immune privilege in corneal transplantation, Prog. Retin. Eye Res 72 (2019) 100758.

Available from: https://doi.org/10.1016/j.preteyeres.2019.
04.002.
[21] J. Bennett, Taking stock of retinal gene therapy: looking back and moving forward, Mol. Ther. 25 (2017) 1076–1094.
[22] J.J. Darrow, Luxturna: FDA documents reveal the value of a costly gene therapy, Drug Discov. Today 24 (4) (2019) 949–954.
[23] A.S. Lewin, B. Rossmiller, H. Mao, Gene augmentation for adRP mutations in RHO, Cold Spring Harb. Perspect. Med. 4 (9) (2014) a017400. Available from: https://doi.org/10.1101/cshperspect.a017400.
[24] M.M. LaVail, D. Yasumura, M.T. Matthes, K.A. Drenser, J.G. Flannery, A.S. Lewin, et al., Ribozyme rescue of photoreceptor cells in P23H transgenic rats: long term survival and late-stage therapy, Proc. Natl Acad. Sci. U.S.A. 97 (2000) 11488–11493.
[25] M. Ader, P.F. Kenna, S. Millington-Ward, G. Clark, A. Kennan, M. O'Reilly, et al., Toward a gene therapy for dominant disease: validation of an RNA interference-based mutation-independent approach, Mol. Ther. 12 (2005) 555–561.
[26] S. Millington-Ward, N. Chadderton, M. O'Reilly, A. Palfi, T. Goldmann, C. Kilty, et al., Suppression and replacement gene therapy for autosomal dominant disease in a murine model of dominant retinitis pigmentosa, Mol. Ther. 19 (2011) 642–649.
[27] E.H. Allen, D.G. Courtney, S.D. Atkinson, J.E. Moore, L. Mairs, E.T. Poulsen, et al., Keratin 12 missense mutation induces the unfolded protein response and apoptosis in Meesmann epithelial corneal dystrophy, Hum. Mol. Genet. 25 (2016) 1176–1191.
[28] D.G. Courtney, E.T. Poulsen, S. Kennedy, J.E. Moore, S.D. Atkinson, E. Maurizi, et al., Protein composition of TGFBI-R124C- and TGFBI-R555W-associated aggregates suggests multiple mechanisms leading to lattice and granular corneal dystrophy, Invest. Ophthalmol. Vis. Sci. 56 (2015) 4653–4661.
[29] K.A. Christie, D.G. Courtney, L.A. DeDionisio, C. Chao-Shern, S. De Majumdar, L.C. Mairs, et al., Towards personalised allele-specific CRISPR gene editing to treat autosomal dominant disorders, Sci. Rep. 7 (2017) 16174.
[30] T. Sakimoto, H. Kanno, J. Shoji, Y. Kashima, S. Nakagawa, S. Miwa, et al., A novel nonsense mutation with a compound heterozygous mutation in TGFBI gene in lattice corneal dystrophy type I, Jpn. J. Ophthalmol. 47 (2003) 13–17.
[31] J.S. Song, D.H. Lim, E.S. Chung, T.Y. Chung, C.S. Ki, Mutation analysis of the TGFBI gene in consecutive Korean patients with corneal dystrophies, Ann. Lab. Med. 35 (2015) 336–340.
[32] S.K. Ahlfeld, J. Wang, Y. Gao, P. Snider, S.J. Conway, Initial suppression of transforming growth factor-β signaling and loss of TGFBI causes early alveolar structural defects resulting in bronchopulmonary dysplasia, Am. J. Pathol. 186 (4) (2016) 777–793.
[33] W.W. Kao, C.Y. Liu, R.L. Converse, A. Shiraishi, C.W. Kao, M. Ishizaki, et al., Keratin 12-deficient mice have fragile corneal epithelia, Invest. Ophthalmol. Vis. Sci. 37 (1996) 2572–2584.
[34] E.T. Poulsen, K. Runager, N.S. Nielsen, M.V. Lukassen, K. Thomsen, P. Snider, et al., Proteomic profiling of TGFBI-null mouse corneas reveals only minor changes in matrix composition supportive of TGFBI knockdown as therapy against TGFBI-linked corneal dystrophies, FEBS J. 285 (2018) 101–114.
[35] D.G. Courtney, S.D. Atkinson, J.E. Moore, E. Maurizi, C. Serafini, G. Pellegrini, et al., Development of allele-specific gene-silencing siRNAs for TGFBI Arg124Cys in lattice corneal dystrophy type I, Invest. Ophthalmol. Vis. Sci. 55 (2014) 977–985.
[36] D.G. Courtney, S.D. Atkinson, E.H. Allen, J.E. Moore, C.P. Walsh, D.M. Pedrioli, et al., siRNA silencing of the mutant keratin 12 allele in corneal limbal epithelial cells grown from patients with Meesmann's epithelial corneal dystrophy, Invest. Ophthalmol. Vis. Sci. 55 (2014) 3352–3360.
[37] E.H.A. Allen, S.D. Atkinson, H. Liao, J.E. Moore, D.M. L. Pedrioli, F.J.D. Smith, et al., Allele-specific siRNA silencing for the common keratin 12 founder mutation in Meesmann epithelial corneal dystrophy, Invest. Ophthalmol. Vis. Sci. 54 (2013) 494–502.
[38] H. Liao, A.D. Irvine, C.J. Macewen, K.H. Weed, L. Porter, L.D. Corden, et al., Development of allele-specific therapeutic siRNA in Meesmann epithelial corneal dystrophy, PLoS One 6 (2011) e28582.
[39] W.H. McLean, C.B. Moore, Keratin disorders: from gene to therapy, Hum. Mol. Genet. 20 (R2) (2011) R189–R197.
[40] V.S. Yellore, S.A. Rayner, A.J. Aldave, TGFB1-induced extracellular expression of TGFBIp and inhibition of TGFBIp expression by RNA interference in a human corneal epithelial cell line, Invest. Ophthalmol. Vis. Sci. 52 (2011) 757–763.
[41] C. Yuan, E.J. Zins, A.F. Clark, A.J. Huang, Suppression of keratoepithelin and myocilin by small interfering RNAs (siRNA) in vitro, Mol. Vis. 13 (2007) 2083–2095.
[42] D. Schiroli, M.J. Gómara, E. Maurizi, S.D. Atkinson, L. Mairs, K.A. Christie, et al., Effective in vivo topical delivery of siRNA and gene silencing in intact corneal epithelium using a modified cell penetrating peptide, Mol. Ther. Nucleic Acid 17 (2019) 891–906.

[43] F.A. Ran, P.D. Hsu, J. Wright, V. Agarwala, D.A. Scott, F. Zhang, Genome engineering using the CRISPR-Cas9 system, Nat. Protoc. 8 (2013) 2281–2308.
[44] T. Iyama, D.M. Wilson, DNA repair mechanisms in dividing and non-dividing cells, DNA Repair (Amst.) 12 (2013) 620–636.
[45] Y. Taketani, K. Kitamoto, T. Sakisaka, M. Kimakura, T. Toyono, S. Yamagami, et al., Repair of the TGFBI gene in human corneal keratocytes derived from a granular corneal dystrophy patient via CRISPR/Cas9-induced homology-directed repair, Sci. Rep. 7 (2017) 16713.
[46] E.K. Kim, S. Kim, Y.S. Maeng, Generation of TGFBI knockout ABCG2 + /ABCB5 + double-positive limbal epithelial stem cells by CRISPR/Cas9-mediated genome editing, PLoS One 14 (2019) e0211864.
[47] R. Fernandez-Buenaga, F. Aiello, S.S. Zaher, A. Grixti, S. Ahmad, Twenty years of limbal epithelial therapy: an update on managing limbal stem cell deficiency, BMJ Open Ophthalmol. 3 (2018) e000164.
[48] M. Jinek, K. Chylinski, I. Fonfara, M. Hauer, J.A. Doudna, E. Charpentier, A programmable dual-RNA – guided DNA endonuclease in adaptive bacterial immunity, Science 337 (2012) 816–822.
[49] F.J.M. Mojica, C. Díez-Villaseñor, J. García-Martínez, C. Almendros, Short motif sequences determine the targets of the prokaryotic CRISPR defence system, Microbiology 155 (2009) 733–740.
[50] D.G. Courtney, J.E. Moore, S.D. Atkinson, E. Maurizi, E.H.A. Allen, D.M.L. Pedrioli, et al., CRISPR/Cas9 DNA cleavage at SNP-derived PAM enables both in vitro and in vivo KRT12 mutation-specific targeting, Gene Ther. 23 (2016) 108–112.
[51] K. Fujiki, Y. Hotta, K. Nakayasu, T. Yokoyama, T. Takano, T. Yamaguchi, et al., A new L527R mutation of the betaIGH3 gene in patients with lattice corneal dystrophy with deep stromal opacities, Hum. Genet. 103 (1998) 286–289.
[52] R. Cencic, H. Miura, A. Malina, F. Robert, S. Ethier, T. M. Schmeing, et al., Protospacer adjacent motif (PAM)-distal sequences engage CRISPR Cas9 DNA target cleavage, PLoS One 9 (2014) e109213.
[53] C. Anders, O. Niewoehner, M. Jinek, In vitro reconstitution and crystallization of Cas9 endonuclease bound to a guide RNA and a DNA target, Methods Enzymol. 558 (2015) 515–537.
[54] S.H. Sternberg, S. Redding, M. Jinek, E.C. Greene, J. A. Doudna, DNA interrogation by the CRISPR RNA-guided endonuclease Cas9, Nature 507 (2014) 62–67.
[55] V. Pattanayak, S. Lin, J.P. Guilinger, E. Ma, J.A. Doudna, D.R. Liu, High-throughput profiling of off-target DNA cleavage reveals RNA-programmed Cas9 nuclease specificity, Nat. Biotechnol. 31 (2013) 839–843.
[56] X. Wu, D.A. Scott, A.J. Kriz, A.C. Chiu, P.D. Hsu, D.B. Dadon, et al., Genome-wide binding of the CRISPR endonuclease Cas9 in mammalian cells, Nat. Biotechnol. 32 (2014) 670–676.
[57] S. Lessard, L. Francioli, J. Alfoldi, J.C. Tardif, P.T. Ellinor, D.G. MacArthur, et al., Human genetic variation alters CRISPR-Cas9 on- and off-targeting specificity at therapeutically implicated loci, Proc. Natl Acad. Sci. U.S.A. 114 (2017) E11257–E11266.
[58] D.A. Scott, F. Zhang, Implications of human genetic variation in CRISPR-based therapeutic genome editing, Nat. Med. 23 (2017) 1095–1101.
[59] J.M. Lee, K.H. Kim, A. Shin, M.J. Chao, K. Abu Elneel, T. Gillis, et al., Sequence-level analysis of the major European Huntington disease haplotype, Am. J. Hum. Genet. 97 (2015) 435–444.
[60] J.W. Shin, K.H. Kim, M.J. Chao, R.S. Atwal, T. Gillis, M.E. MacDonald, J.F. Gusella, J.M Lee, Permanent inactivation of Huntington's disease mutation by personalized allele-specific CRISPR/Cas9, Hum. Mol. Genet. 25 (20) (2016) 4566–4576. Available from: https://doi.org/10.1093/hmg/ddw286.
[61] G.X. Zheng, B.T. Lau, M. Schnall-Levin, M. Jarosz, J. M. Bell, C.M. Hindson, et al., Haplotyping germline and cancer genomes with high-throughput linked-read sequencing, Nat. Biotechnol. 34 (2016) 303–311.
[62] G.X. Ruan, E. Barry, D. Yu, M. Lukason, S.H. Cheng, A. Scaria, CRISPR/Cas9-mediated genome editing as a therapeutic approach for Leber congenital amaurosis 10, Mol. Ther. 25 (2017) 331–341.
[63] Q. Zheng, X. Cai, M.H. Tan, S. Schaffert, C.P. Arnold, X. Gong, et al., Precise gene deletion and replacement using the CRISPR/Cas9 system in human cells, Biotechniques 57 (2014) 115–124.
[64] E. Nagy, L.E. Maquat, A rule for termination-codon position within intron-containing genes: when nonsense affects RNA abundance, Trends Biochem. Sci. 23 (1998) 198–199.
[65] M.W. Popp, L.E. Maquat, Leveraging rules of nonsense-mediated mRNA decay for genome engineering and personalized medicine, Cell 165 (2016) 1319–1322.
[66] B. Yavuz, U.B. Kompella, Ocular drug delivery, Handb. Exp. Pharmacol. 242 (2017) 57–93.
[67] A. Guzman-Aranguez, P. Loma, J. Pintor, Small-interfering RNAs (siRNAs) as a promising tool for ocular therapy, Br. J. Pharmacol. 170 (2013) 730–747.
[68] A. Subrizi, E.M. Del Amo, V. Korzhikov-Vlakh, T. Tennikova, M. Ruponen, A. Urtti, Design principles of ocular drug delivery systems: importance of drug payload, release rate, and material properties, Drug Discov. Today 24(8) (2019) 1446–1457.
[69] Y.C. Kim, B. Chiang, X. Wu, M.R. Prausnitz, Ocular delivery of macromolecules, J. Control. Release 190 (2014) 172–181.

[70] K.M. Hämäläinen, K. Kananen, S. Auriola, K. Kontturi, A. Urtti, Characterization of paracellular and aqueous penetration routes in cornea, conjunctiva, and sclera, Invest. Ophthalmol. Vis. Sci. 38 (1997) 627–634.
[71] A.V. Oliveira, A.M. Rosa da Costa, G.A. Silva, Non-viral strategies for ocular gene delivery, Mater. Sci. Eng. C Mater. Biol. Appl 77 (2017) 1275–1289.
[72] Z. Li, F. Duan, L. Lin, Q. Huang, K. Wu, A new approach of delivering siRNA to the cornea and its application for inhibiting herpes simplex keratitis, Curr. Mol. Med. 14 (2014) 1215–1225.
[73] J.G. Souza, K. Dias, T.A. Pereira, D.S. Bernardi, R.F. Lopez, Topical delivery of ocular therapeutics: carrier systems and physical methods, J. Pharm. Pharmacol. 66 (2014) 507–530.
[74] M. Matthaei, H. Meng, I. Bhutto, Q. Xu, E. Boelke, J. Hanes, et al., Systematic assessment of microneedle injection into the mouse cornea, Eur. J. Med. Res. 17 (2012) 19.
[75] R.R. Mohan, A. Sharma, M.V. Netto, S. Sinha, S.E. Wilson, Gene therapy in the cornea, Prog. Retin. Eye Res. 24 (2005) 537–559.
[76] R.R. Mohan, J.C.K. Tovey, A. Sharma, A. Tandon, Gene therapy in the cornea: 2005-present, Prog. Retin. Eye Res. 31 (2012) 43–64.
[77] T. Borrás, B.T. Gabelt, G.K. Klintworth, J.C. Peterson, P.L. Kaufman, Non-invasive observation of repeated adenoviral GFP gene delivery to the anterior segment of the monkey eye in vivo, J. Gene Med. 3 (2001) 437–449.
[78] K. Tsubota, H. Inoue, K. Ando, M. Ono, K. Yoshino, I. Saito, Adenovirus-mediated gene transfer to the ocular surface epithelium, Exp. Eye Res. 67 (1998) 531–538.
[79] A. Sharma, J.C.K. Tovey, A. Ghosh, R.R. Mohan, AAV serotype influences gene transfer in corneal stroma in vivo, Exp. Eye Res. 91 (2010) 440–448.
[80] E. Zinn, L.H. Vandenberghe, Adeno-associated virus: fit to serve, Curr. Opin. Virol. 8 (2014) 90–97.
[81] T. Athanasopoulos, M.M. Munye, R.J. Yáñez-Muñoz, Nonintegrating gene therapy vectors, Hematol. Oncol. Clin. N. Am. 31 (2017) 753–770.
[82] T. Borrás, C.R. Brandt, R. Nickells, R. Ritch, Gene therapy for glaucoma: treating a multifaceted, chronic disease, Invest Ophthalmol Vis Sci 43 (2002) 2513–2518. 12147578.
[83] C. Hippert, S. Ibanes, N. Serratrice, F. Court, F. Malecaze, E.J. Kremer, et al., Corneal transduction by intra-stromal injection of AAV vectors in vivo in the mouse and ex vivo in human explants, PLoS One 7 (2012) e35318.
[84] J. O'Callaghan, D.E. Crosbie, P.S. Cassidy, J.M. Sherwood, C. Flügel-Koch, E. Lütjen-Drecoll, et al., Therapeutic potential of AAV-mediated MMP-3 secretion from corneal endothelium in treating glaucoma, Hum. Mol. Genet. 26 (2017) 1230–1246.
[85] M. Basche, D. Kampik, S. Kawasaki, M.J. Branch, M. Robinson, D.F. Larkin, et al., Sustained and widespread gene delivery to the corneal epithelium via in situ transduction of limbal epithelial stem cells, using lentiviral and adeno-associated viral vectors, Hum. Gene Ther. 29 (2018) 1140–1152.

CHAPTER

8

Employing nonhomologous end joining and homology-directed repair for treatment of Leber congenital amaurosis and inherited retinal degeneration

Dong Hyun Jo[4], Jin Hyoung Kim[1] and Jeong Hun Kim[1,2,3]

[1]Fight against Angiogenesis-Related Blindness (FARB) Laboratory, Biomedical Research Institute, Seoul National University Hospital, Seoul, Republic of Korea [2]Department of Biomedical Sciences, Seoul National University College of Medicine, Seoul, Republic of Korea [3]Department of Ophthalmology, Seoul National University College of Medicine, Seoul, Republic of Korea [4]Department of Anatomy and Cell Biology, Seoul National University College of Medicine, Seoul, Republic of Korea

Leber congenital amaurosis and inherited retinal degeneration

Inherited retinal degeneration (IRD) indicates a group of retinal diseases caused by gene mutations affecting the physiologic function of photoreceptors, retinal pigment epithelium (RPE), or inner retina [1]. IRD includes retinitis pigmentosa (RP), cone and cone–rod dystrophy (CORD), Leber congenital amaurosis (LCA), congenital stationary night blindness, Usher syndrome, Bardet–Biedl syndrome, Joubert syndrome, Stargardt disease, and achromatopsia. Patients with IRDs usually experience progressive visual loss. At a molecular level, IRDs result from pathogenic mutations from single gene [2]. Mutations in more than 200 genes which are related to the function of the visual system are associated with IRD [3,4].

LCA type 2 (LCA2) is the most widely studied target disease among IRD. Based on studies on mice, beagles, nonhuman primates (NHP), and clinical trials [5–13], gene therapy for *RPE65* delivery comes to the clinic using adeno-associated virus (AAV).

Genetic aberrations in patients with LCA

LCA is characterized by severe visual impairment, photophobia, and nystagmus, caused by mutations in genes involved in the visual

CRISPR Genome Surgery in Stem Cells and Disease Tissues
DOI: https://doi.org/10.1016/B978-0-12-817876-8.00001-2

cycle [14]. LCA is divided into 18 types according to the genes in which mutations occur. Specific genes in which mutations occur according to each type of LCA are summarized in Table 8.1.

Genetic aberrations in patients with other IRDs

Other IRDs are also induced by mutations in genes which are related with functions of photoreceptors and RPE. Among them, RP is characterized by night blindness and loss of peripheral vision [15]. Specific genes in which mutations occur in each IRD are summarized in Table 8.2.

Genome surgery for treatment of human diseases

So far, there have been no direct treatment options to treat patients with LCA and other IRDs besides gene therapy for LCA2. Genome surgery provides several advantages in the treatment of LCA and other IRDs (Fig. 8.1). First, the mutated sequence can be totally exchanged with the normal sequence, unlike gene therapy in which the mutated sequence remains with the

TABLE 8.1 Genetic aberrations in patients with LCA.

Type	Gene	Location	Phenotype MIM number	Inheritance pattern
LCA1	GUCY2D	17p13.1	204000	AR
LCA2	RPE65	1p31.3	204100	AR
LCA3	SPATA7	14q31.3	604232	N/S
LCA4	AIPL1	17p13.2	604393	AR
LCA5	LCA5	6q14.1	604573	N/S
LCA6	RPGRIP1	14q11.2	613826	N/S
LCA7	CRX	19q13.33	613829	N/S
LCA8	CRB1	1q31.3	613835	N/S
LCA9	NMNAT1	1p36.22	608553	AR
LCA10	CEP290	12q21.32	611755	N/S
LCA11	IMPDH1	7q32.1	613837	AD
LCA12	RD3	1q32.3	610612	AR
LCA13	RDH12	14q24.1	612712	AD, AR
LCA14	LRAT	4q32.1	613341	AR
LCA15	TULP1	6p21.31	613843	AR
LCA16	KCNJ13	2q37.1	614186	AR
LCA17	GDF6	8q22.1	615360	AR
LCA18	PRPH2	6p21.1	608133	AD, AR

Notes: AD, Autosomal dominant; *AR*, autosomal recessive; *N/S*, not specified.
Data from Online Mendelian Inheritance in Man (OMIM) Database.

TABLE 8.2 Genetic aberrations in patients with IRD.

Disease	Genes	Phenotypic series number	Inheritance pattern
RP	ABCA4, AGBL5, AHR, ARHGEF18, ARL2BP, ARL3, ARL6, BBS2, BEST1, C8orf37, CA4, CDHR1, CERKL, CLRN1, CNGB1, CRB1, DHDDS, DHX38, EYS, FAM161A, FSCN2, GUCA1B, FSCN2, HGSNAT, HK1, IDH3B, IFT140, IFT172, IF43, IMPDH1, IMPG2, KLHL7, KIZ, MAK, MERTK, NEK2, NR2E3, NRL, OFD1, PCARE, PDE6A, PDE6G, POMGNT1, PRCD, PRPF3, PRPF31, PRPF4, PRPF6, PRPF8, PRPH2, RBP3, RDH12, REEP6, RGR, RHO, ROM1, RP1, RP2, RP22, RP24, RP32, RP34, RP6, RP63, RP9, RPE65, RPGR, RPY, SAG, SEMA4A, SLC7A14, SNRNP200, TOPORS, TULP1, TTC8, USH2A, ZNF408, ZNF513	PS180100	AD, AR, XL, XLR
Cone–rod dystrophy	ABCA4, ADAM9, C8orf37, CACNA1F, CDHR1, COD2, CORD1, CORD17, CORD8, CRX, DRAM2, GUCA1A, GUCY2D, OPN1LW, OPN1MW, PDE6C, POC1B, PITPNM3, PROM1, RAB28, RAX2, RIMS1, RPGR, RPGRIP1, SEMA4A, TTLL5	PS120970	AD, AR, XL, XLR
Congenital stationary night blindness	CACNA1F, GNAT1, GNB3, GPR179, GRK1, GRM6, LRIT3, NYX, PDE6B, RHO, SAG, SLC24A1, TRPM1	PS310500	AD, AR, XL, XLR
Usher syndrome	ADGRV1, ARSG, CDH23, CIB2, CLRN1, HARS, MYO7A, PCDH15, PDZD7, SANS, USH1C, USH1E, USH1H, USH1K, USH2A, WHRN	PS276900	AR, DR
Bardet–Biedl syndrome	ARL6, BBIP1, BBS1, BBS10, BBS12, BBS2, BBS4, BBS5, BBS7, C8orf37, CCDC28B, CEP290, IFT27, IFT74, LZTFL1, MKKS, MKS1, PTHB1, SDCCAG8, TMEM67, TRIM32, TTC8, WDPCP	PS209900	AR, DR
Joubert syndrome	AHI1, ARL13B, ARL3, ARMC9, B9D1, B9D2, CC2D2A, CEP104, CEP120, CEP290, CEP41, CPLANE1, CSPP1, INPP5E, KATNIP, KIAA0586, KIF7, MKS1, NPHP1, OFD1, PDE6D, PIBF1, RPGRIP1L, SUFU, TCTN2, TCTN3, TECT1, TMEM107, TMEM138, TMEM216, TMEM231, TMEM237, TMEM67, ZNF423	PS213300	AD, AR, XLR
Stargardt disease	ABCA4, ELOVL4, PROM1	N/A	AD, AR
Achromatopsia	ATF6, CNGA3, CNGB3, GNAT2	N/A	AR

Notes: AD, Autosomal dominant; *AR,* autosomal recessive; *Dr,* digenic recessive; *N/A,* not available; *XL,* X-linked; *XLR,* X-linked recessive.
Data from Online Mendelian Inheritance in Man (OMIM) Database.

adopted normal sequence. Second, genome surgery corrects the disease condition even when the size of the gene is larger than the loading capacity of viral vectors. Third, the pathogenic mutations that produce the disease-causing protein can be fixed to decrease the production. Fourth, when the desired DNA sequence is incorporated into the targeted genome, the cells regulate the expression of genes through the normal gene-regulatory machinery.

In the application of genome surgery, one of the most critical concerns is the selection of delivery tools. In large, there are two types of therapeutic approaches of genome surgery: ex vivo and in vivo [16]. In the ex vivo approach, the clustered regularly interspaced short palindromic repeats (CRISPR) system can be applied via electroporation or viral vectors. Alternatively, the in vivo approach delivered the CRISPR system to target organs through the

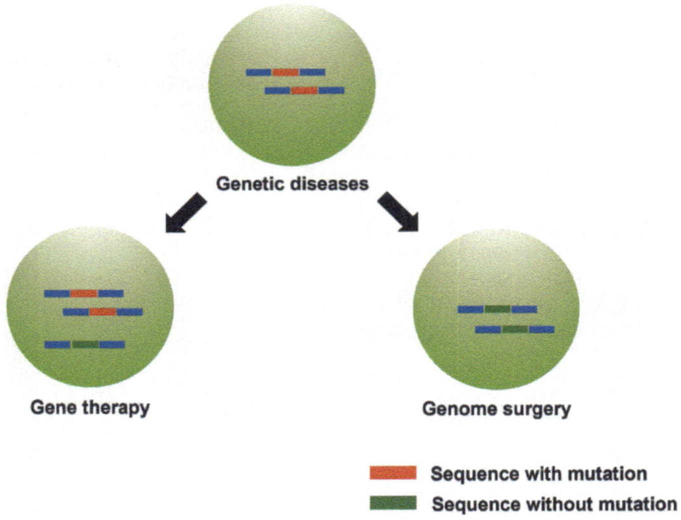

FIGURE 8.1 Gene therapy and genome surgery. Gene therapy restores the wild-type gene with a remaining sequence containing mutations. On the other hand, genome surgery, if possible, totally corrects the sequence with mutations with the wild-type sequence.

bloodstream or directly. The eye is an appropriate target for innovative treatment options including gene and cell therapy. There are well-established protocols for intravitreal and subretinal administration based on several clinical trials on AAV-mediated gene delivery [8–13] and implantation of RPE cells from embryonic or induced pluripotent stem cells [17,18]. It is anticipated that genome surgery for treatment of ocular diseases can be performed based on these local administration routes.

Modes of correction with genome surgery

In this chapter, genome surgery indicated the correction of genes by the CRISPR/CRISPR-associated protein (Cas) system, which is a ribonucleoprotein consisting of the Cas protein and two RNA components, CRISPR-RNA (crRNA) and transactivating crRNA (tracrRNA) [19]. crRNA and tracrRNA can be combined to single-guide RNA (sgRNA) for the application. Unlike previous generations of genome editing such as zinc-finger nucleases and transcription activator-like effector nucleases which are guided by protein–DNA binding [20], CRISPR/Cas9 uses Watson–Crick base pairing between the CRISPR-associated RNA and the target DNA for targeting the specific sites in the genome. One of the most widely utilized Cas9, Cas9 from *Streptococcus pneumoniae* (SpCas9), induces the double-stranded break (DSB) at three base upstream of the protospacer adjacent motif (PAM), NGG (N indicates any nucleotide), in the target genome sequence [21]. The VQR variant of SpCas9 recognizes the NGA PAM [22]. *Staphylococcus aureus* Cas9 (SaCas9) and *Campylobacter jejuni* Cas9 use the PAM of NNGRRT (R indicates the purine nucleotides A or G) [22] and NNNNACAC or NNNNRYAC (Y indicates the pyrimidine nucleotides C or T) [23], respectively. Furthermore, there are ongoing efforts to develop novel orthologs of the Cas9 system to use different PAMs for further coverage of the target sites.

Upon the DSB by different orthologs of Cas9, it induces the cell-intrinsic repair mechanisms in two distinct patterns, nonhomologous end joining (NHEJ) and homology-directed repair (HDR) [24]. These two pathways are highly conserved to repair DSBs [25–27]. Genome surgery uses NHEJ and HDR for the removal or correction of pathogenic mutations (Fig. 8.2).

Nonhomologous end joining

NHEJ is a repair mechanism of DSB by directly rejoining the two cleaved ends of the target gene [28]. During this process, small insertions or deletions (indels) of various sizes occur to generate frameshift mutations, which lead to mRNA degradation or production of nonfunctional truncated proteins [29]. In this context, NHEJ can be used to remove gain-of-function or dominant-negative mutations [30]. Interestingly, the patterns of indels are quite preserved according to the genome surgery machinery [31].

Homology-directed repair

On the other hand, in the presence of the homologous DNA sequence or the donor DNA template, HDR occurs to repair the DSB. In this case, the targeted sequence is restored to the wild-type sequence. Although HDR is a tool to correct point mutations, it is widely believed that HDR is restricted to the S and G2 phase of the cell cycle in dividing cells [32]. The cell cycle phase and cyclin-dependent kinases play roles in the mode of correction upon the DSB [33]. For effective HDR using genome surgery, the DSB site should be as close to the insertion site for the mutation [4], and the homology arms of the donor template should be long enough [34].

Even with the successful cases with gene therapy, the HDR using genome surgery has a

```
            Trp Leu Thr Gly Ser Leu Leu Arg Cys Gly Pro Gly Leu Phe Glu
WT      5′-TGG CTC ACT GGC AGT CTC CTC CGA TGT GGG CCA GGG CTC TTT GAA-3′
        3′-ACC GAG TGA CCG TCA GAG GAG GCT ACA CCC GGT CCC GAG AAA CTT-5′

            Trp Leu Thr Gly Ser Leu Leu  X   -   -   -   -   -   -   -
MT      5′-TGG CTC ACT GGC AGT CTC CTC TGA TGT GGG CCA GGG CTC TTT GAA-3′
        3′-ACC GAG TGA CCG TCA GAG GAG ACT ACA CCC GGT CCC GAG AAA CTT-5′

            Trp Leu Thr Gly Ser Leu Leu  -  Cys Gly Pro Gly Leu Phe Glu
NHEJ-1  5′-TGG CTC ACT GGC AGT CTC CTC --- TGT GGG CCA GGG CTC TTT GAA-3′
        3′-ACC GAG TGA CCG TCA GAG GAG --- ACA CCC GGT CCC GAG AAA CTT-5′

NHEJ-2  5′-TGG CTC ACT GGC AGT CTC CT- --A TGT GGG CCA GGG CTC TTT GAA-3′
        3′-ACC GAG TGA CCG TCA GAG GA- --T ACA CCC GGT CCC GAG AAA CTT-5′

            Trp Leu Thr Gly Ser Leu Leu Arg Cys Gly Pro Gly Leu Phe Glu
HDR-1   5′-TGG CTC ACT GGC AGT CTC CTC CGA TGT GGG CCA GGG CTC TTT GAA-3′
        3′-ACC GAG TGA CCG TCA GAG GAG GCT ACA CCC GGT CCC GAG AAA CTT-5′

HDR-2   5′-TGG CTC ACT GGC AGT CTC CTC CGT TGT GGG CCA GGG CTC TTT GAA-3′
        3′-ACC GAG TGA CCG TCA GAG GAG GCA ACA CCC GGT CCC GAG AAA CTT-5′
```

FIGURE 8.2 Examples of two modes of correction after DSB, NHEJ versus HDR. A mutation in the *RPE65* gene in *rd12* mice can be corrected by two distinct pathways, NHEJ and HDR. Amino acid sequences are indicated above the DNA sequences. Red bases in the MT sequence indicate the converted ones from the WT sequence. Azure bases in the HDR-corrected sequences indicate the converted ones to the WT sequence at the levels of DNA or amino acids. *MT*, Mutant sequence; *WT*, wild-type sequence.

role in the treatment of retinal diseases. The capacity of AAV, the most widely utilized viral vector for clinical use, is ~4.8 kB, limiting the potential of the delivery of genes of which sizes are above it. In these cases, HDR can be applied to recover the wild-type sequence by total correction of the mutated sequences. Furthermore, AAVs by themselves stimulate HDR in mammalian cells [35]. The nuclear localization of genomes and the recombinogenic inverted terminal repeats of AAVs are known to affect the efficiency of HDR [36].

Employing NHEJ and HDR for treatment of LCA and IRD

Genome surgery with CRISPR/Cas9 can be used as a tool to remove or correct pathological mutations associated with the pathogenesis of LCA and other IRDs [16]. NHEJ and HDR can be utilized, respectively, for targeted gene disruption and correction [37]. The approaches using NHEJ and HDR in animal models of LCA and other IRDs are summarized in Table 8.3.

Nonhomologous end joining

The most common mutation in patients with LCA type 10 (LCA10) is an adenine to guanine mutation located in the intron 26 (intron variant sequence 26 [IVS26], c. 2991 + 1655A > G) of the *CEP290* gene to create a strong novel splice donor and insert a cryptic exon [43]. The mutation leads to the inclusion of a cryptic exon containing a premature stop codon (p.C998X) into the cellular *CEP290* transcripts [43]. Using an AAV type 5 (AAV5) vector containing SaCas9 and CEP290-specific gRNAs, Maeder et al. induced removal or inversion of the IVS26 mutation, resulting in normal splicing [38]. In this study, the authors utilized knock-in mice with human CEP290 IVS26 [44], showing the editing rate of ~20% through the 6- and 9-month duration of the study with subretinal injection of 1×10^{12} and 1×10^{13} viral genomes (vg)/mL of AAV5 (injection dose: 1 μL) [38]. Further studies with surrogate sgRNAs targeting to NHP demonstrated the editing rate of 16.1% ± 2.8% and 27.9% ± 20.7% in the 7×10^{11} and 1×10^{12} vg/mL dose groups, respectively [38].

In another study, Ruan et al. demonstrated the effective deletion of an intronic fragment of the *Cep290* gene in the murine retina through a dual AAV system containing SpCas9 and sgRNA separately [39]. Subretinal injection of 1×10^{11} vg each resulted in 7.5%–26.4% of genomic deletion at four weeks postinjection. These results showed that this system could be used to treat the IVS26 splicing mutation associated with patients with LCA10.

Mutations in the *GUCY2D* gene account for ~35% of cases with autosomal dominant CORD [45,46]. Dominantly inherited mutations increase the affinity of retinal guanylyl cyclase 1 (retGC1) to guanylate cyclase activating protein-1, leading to overproduction of cyclic guanosine monophosphate [47–50]. To demonstrate the possibility of the disruption of the mutant *GUCY2D* allele, McCullough et al. induced somatic knockout of the gene coding retGC1 in the mouse (*Gucy2e*) and NHP (*GUCY2D*) retina [40]. Subretinal injection of AAV5 vectors containing SaCas9 resulted in 32%–81% of editing rate in mice and 6.8%–20.2% of editing rate in NHP.

Mutations in the *RHO* gene account for ~25% of autosomal dominant RP [51–53]. Among them, c. 68C > A (p.Pro23His, hereafter referred to as P23H) is the most common form of mutation. In a study from Latella et al. [41], subretinally administered plasmids containing two sgRNAs and the SpCas9 transcript effectively removed the pathogenic mutation by NHEJ from *Rho*-knockout mice expressing the human *RHO* gene with the P23H mutation.

Using the VQR variant of SpCas9, Giannelli et al. demonstrated that the mutant *RHO* allele could be targeted with electroporation or

TABLE 8.3 Attempts employing genome surgery to treat animal models of LCA and inherited retinal degeneration.

Target disease	Target gene	Target mutation	Animal models	Delivery tools	Cas9 isoforms	Delivery methods	Concentration	Dose	Mode of correction	Editing rate	References
LCA2	RPE65	c. 130C > T	*rd12* mice	AAV9	SpCas9	Subretinal injection	2×10^{13}–2×10^{14} vg/mL	1 μL	HDR	1.2%	Jo et al. [59]
LCA10	CEP290	IVS26, c. 2991 + 1655A > G	Human CEP290 knock-in mice, NHP	AAV5	SaCas9	Subretinal injection	1×10^{11}–1×10^{13} vg/mL in mice; 1×10^{11}–1×10^{12} vg/mL in NHP	1 μL in mice; 100 μL in NHP	NHEJ	5.5%–60.8% in mice; 3.5%–27.9% in NHP	Maeder et al. [38]
LCA10	CEP290	Intron 25 (homologous to intron 26)	C57BL/6 mice	AAV5	SpCas9	Subretinal injection	2×10^{12} vg/mL	1 μL	NHEJ	7.5%–26.4%	Ruan et al. [39]
CORD6	GUCY2D	WT sequence	C57BL/6 mice, NHP	AAV5	SaCas9	Subretinal injection	3×10^{11}–3×10^{13} vg/mL in mice;	1 μL in mice; 50–120 μL in NHP	NHEJ	32%–81% in mice; 6.8%–20.2% in NHP	McCullough et al. [40]
RP4	RHO	c. 68C > A	Homozygous P23H knock-in mice	Electroporation, AAV2–7m8	SpCas9-VQR	Intravitreal injection	1×10^{12} vg/mL	1 μL	NHEJ	93% with electroporation; 8%–27% with AAV	Giannelli et al. [37]
RP4	RHO	c. 68C > A	$Rho^{-/-}$ P23H knock-in mice	Electroporation	SpCas9	Subretinal injection	6 mg/mL	0.5 μL	NHEJ	4%–33% of GFP-positive cells	Latella et al. [41]
RP4	RHO	p.Ser334X	S334ter rats	Electroporation	SpCas9	Subretinal injection	6.0–7.4 mg/mL	1 μL	NHEJ	33%–36%	Bakondi et al. [42]

Notes: AAV, Adeno-associated virus; *CORD*, cone–rod dystrophy; *LCA*, Leber congenital amaurosis; *NHEJ*, nonhomologous end joining; *NHP*, nonhuman primate; *RP*, retinitis pigmentosa; *SaCas9*, *Staphylococcus aureus* Cas9; *SpCas9*, *Streptococcus pneumoniae* Cas9; *WT*, wild type.

intravitreal injection of AAV in homozygous P23H knock-in mice [37]. The editing rates of AAV containing the SpCas9-VQR variant were ranged from 8% to 27%.

Another animal model of autosomal dominant RP is transgenic S334ter rats which possess the mouse genomic fragment containing a mutation (p.Ser334X) in the *Rho* gene [54,55]. Subretinal injection of the vector containing sgRNA targeting to the mutant sequence and the Cas9 transcript resulted in the allele-specific ablation of the pathogenic mutant in the S334ter rats [42].

Homology-directed repair

It is generally known that the postmitotic state of the cells in the retina and RPE hinders efficient genome surgery, especially HDR [4]. However, recent studies showed that AAV-mediated delivery of CRISPR genome surgery machinery induced HDR even in postmitotic neuron and muscle cells [56,57]. In this context, the HDR-based approaches can be a tool to treat IRDs. In diseases of which pathogenic mutations occur in the genes larger than the packaging ability of viral vectors, the HDR-based approaches could be a realistic alternative to gene augmentation therapy. Wu et al. demonstrated a proof-of-concept in a study using sgRNA, Cas9 protein, and the single-stranded oligodeoxynucleotide donor template for the recovery of the wild-type sequence in the zygotes from *rd1* mice, a model of RP [58]. Jo et al. showed that subretinal delivery of dual AAV vectors carrying Cas9 and donor DNA resulted in ~1.2% of HDR in the RPE of *rd12* mice and recovery of electroretinographic responses [59].

Future direction and conclusions

Genome surgery, if effective, can provide a specific tool to correct pathogenic mutations through NHEJ and HDR. Despite a few successful attempts in animal models of IRDs, there are still issues to be solved in the clinical application of genome surgery for the treatment of IRDs. First, delivery tools (viral and nonviral vectors) and methods (subretinal or intravitreal injection, etc.) should be standardized for broader application with considerable efficiency. Second, researchers, clinicians, and regulatory bodies should meticulously evaluate the risk and benefit of genome surgery not to exaggerate or underestimate them. Then, patients with IRDs would receive the treatment with a favorable risk–benefit ratio. Third, the equality to the therapy should be considered in the application of genome surgery in terms of both accessibility and applicability. Patient groups with low resources or mutations of low incidence should be considered during the development of technologies. If these issues are cleared in the further development of genome surgery, we expect that genome surgery can help patients with IRDs who cannot be treated with conventional treatment options currently available.

References

[1] A. Arbabi, A. Liu, H. Ameri, Gene therapy for inherited retinal degeneration, J. Ocul. Pharmacol. Ther. 35 (2) (2019) 79–97.

[2] P.R. Gupta, R.M. Huckfeldt, Gene therapy for inherited retinal degenerations: initial successes and future challenges, J. Neural Eng. 14 (5) (2017) 051002.

[3] W. Berger, B. Kloeckener-Gruissem, J. Neidhardt, The molecular basis of human retinal and vitreoretinal diseases, Prog. Retin. Eye Res. 29 (5) (2010) 335–375.

[4] M. Yanik, et al., In vivo genome editing as a potential treatment strategy for inherited retinal dystrophies, Prog. Retin. Eye Res. 56 (2017) 1–18.

[5] J. Bennett, et al., Photoreceptor cell rescue in retinal degeneration (rd) mice by in vivo gene therapy, Nat. Med. 2 (6) (1996) 649–654.

[6] G.M. Acland, et al., Gene therapy restores vision in a canine model of childhood blindness, Nat. Genet. 28 (1) (2001) 92–95.

[7] A.V. Cideciyan, et al., Human gene therapy for RPE65 isomerase deficiency activates the retinoid cycle of vision but with slow rod kinetics, Proc. Natl Acad. Sci. U.S.A. 105 (39) (2008) 15112–15117.

[8] J.W. Bainbridge, et al., Effect of gene therapy on visual function in Leber's congenital amaurosis, N. Engl. J. Med. 358 (21) (2008) 2231–2239.
[9] A.M. Maguire, et al., Safety and efficacy of gene transfer for Leber's congenital amaurosis, N. Engl. J. Med. 358 (21) (2008) 2240–2248.
[10] J.W. Bainbridge, et al., Long-term effect of gene therapy on Leber's congenital amaurosis, N. Engl. J. Med. 372 (20) (2015) 1887–1897.
[11] W.W. Hauswirth, et al., Treatment of leber congenital amaurosis due to RPE65 mutations by ocular subretinal injection of adeno-associated virus gene vector: short-term results of a phase I trial, Hum. Gene Ther. 19 (10) (2008) 979–990.
[12] S.G. Jacobson, et al., Improvement and decline in vision with gene therapy in childhood blindness, N. Engl. J. Med. 372 (20) (2015) 1920–1926.
[13] F. Testa, et al., Three-year follow-up after unilateral subretinal delivery of adeno-associated virus in patients with Leber congenital Amaurosis type 2, Ophthalmology 120 (6) (2013) 1283–1291.
[14] A.I. den Hollander, et al., Leber congenital amaurosis: genes, proteins and disease mechanisms, Prog. Retin. Eye Res. 27 (4) (2008) 391–419.
[15] S.P. Daiger, S.J. Bowne, L.S. Sullivan, Perspective on genes and mutations causing retinitis pigmentosa, Arch. Ophthalmol. 125 (2) (2007) 151–158.
[16] D.H. Jo, J.H. Kim, Application of genome engineering for treatment of retinal diseases, BMB Rep. 51 (7) (2018) 315–316.
[17] M. Mandai, et al., Autologous induced stem-cell-derived retinal cells for macular degeneration, N. Engl. J. Med. 376 (11) (2017) 1038–1046.
[18] S.D. Schwartz, et al., Human embryonic stem cell-derived retinal pigment epithelium in patients with age-related macular degeneration and Stargardt's macular dystrophy: follow-up of two open-label phase 1/2 studies, Lancet 385 (9967) (2015) 509–516.
[19] H. Kim, J.S. Kim, A guide to genome engineering with programmable nucleases, Nat. Rev. Genet. 15 (5) (2014) 321–334.
[20] R. Barrangou, J.A. Doudna, Applications of CRISPR technologies in research and beyond, Nat. Biotechnol. 34 (9) (2016) 933–941.
[21] M. Jinek, et al., A programmable dual-RNA-guided DNA endonuclease in adaptive bacterial immunity, Science 337 (6096) (2012) 816–821.
[22] B.P. Kleinstiver, et al., Engineered CRISPR-Cas9 nucleases with altered PAM specificities, Nature 523 (7561) (2015) 481–485.
[23] E. Kim, et al., In vivo genome editing with a small Cas9 orthologue derived from *Campylobacter jejuni*, Nat. Commun. 8 (2017) 14500.
[24] L.S. Symington, J. Gautier, Double-strand break end resection and repair pathway choice, Annu. Rev. Genet. 45 (2011) 247–271.
[25] J. Fishman-Lobell, N. Rudin, J.E. Haber, Two alternative pathways of double-strand break repair that are kinetically separable and independently modulated, Mol. Cell Biol. 12 (3) (1992) 1292–1303.
[26] N. Rudin, E. Sugarman, J.E. Haber, Genetic and physical analysis of double-strand break repair and recombination in *Saccharomyces cerevisiae*, Genetics 122 (3) (1989) 519–534.
[27] F. Liang, et al., Homology-directed repair is a major double-strand break repair pathway in mammalian cells, Proc. Natl Acad. Sci. U.S.A. 95 (9) (1998) 5172–5177.
[28] F.A. Ran, et al., Double nicking by RNA-guided CRISPR Cas9 for enhanced genome editing specificity, Cell 154 (6) (2013) 1380–1389.
[29] M.W. Hentze, A.E. Kulozik, A perfect message: RNA surveillance and nonsense-mediated decay, Cell 96 (3) (1999) 307–310.
[30] C.B.T. Moore, et al., Personalised genome editing - the future for corneal dystrophies, Prog. Retin. Eye Res. 65 (2018) 147–165.
[31] D.H. Jo, et al., Long-term effects of in vivo genome editing in the mouse retina using *Campylobacter jejuni* Cas9 expressed via adeno-associated virus, Mol. Ther. 27 (1) (2019) 130–136.
[32] T. Iyama, D.M. Wilson 3rd, DNA repair mechanisms in dividing and non-dividing cells, DNA Repair. (Amst.) 12 (8) (2013) 620–636.
[33] R. Ceccaldi, B. Rondinelli, A.D. D'Andrea, Repair pathway choices and consequences at the double-strand break, Trends Cell Biol. 26 (1) (2016) 52–64.
[34] K. Li, et al., Optimization of genome engineering approaches with the CRISPR/Cas9 system, PLoS One 9 (8) (2014) e105779.
[35] T. Gaj, B.E. Epstein, D.V. Schaffer, Genome engineering using adeno-associated virus: basic and clinical research applications, Mol. Ther. 24 (3) (2016) 458–464.
[36] R.K. Hirata, D.W. Russell, Design and packaging of adeno-associated virus gene targeting vectors, J. Virol. 74 (10) (2000) 4612–4620.
[37] S.G. Giannelli, et al., Cas9/sgRNA selective targeting of the P23H Rhodopsin mutant allele for treating retinitis pigmentosa by intravitreal AAV9.PHP.B-based delivery, Hum. Mol. Genet. 27 (5) (2018) 761–779.
[38] M.L. Maeder, et al., Development of a gene-editing approach to restore vision loss in Leber congenital amaurosis type 10, Nat. Med. 25 (2) (2019) 229–233.
[39] G.X. Ruan, et al., CRISPR/Cas9-mediated genome editing as a therapeutic approach for Leber congenital amaurosis 10, Mol. Ther. 25 (2) (2017) 331–341.

[40] K.T. McCullough, et al., Somatic gene editing of GUCY2D by AAV-CRISPR/Cas9 alters retinal structure and function in mouse and macaque, Hum. Gene Ther. 30 (5) (2018) 571–589.
[41] M.C. Latella, et al., In vivo editing of the human mutant rhodopsin gene by electroporation of plasmid-based CRISPR/Cas9 in the mouse retina, Mol. Ther. Nucleic Acids 5 (11) (2016) e389.
[42] B. Bakondi, et al., In vivo CRISPR/Cas9 gene editing corrects retinal dystrophy in the S334ter-3 rat model of autosomal dominant retinitis pigmentosa, Mol. Ther. 24 (3) (2016) 556–563.
[43] A.I. den Hollander, et al., Mutations in the CEP290 (NPHP6) gene are a frequent cause of Leber congenital amaurosis, Am. J. Hum. Genet. 79 (3) (2006) 556–561.
[44] A. Garanto, et al., Unexpected CEP290 mRNA splicing in a humanized knock-in mouse model for Leber congenital amaurosis, PLoS One 8 (11) (2013) e79369.
[45] A.M. Payne, et al., Clustering and frequency of mutations in the retinal guanylate cyclase (GUCY2D) gene in patients with dominant cone-rod dystrophies, J. Med. Genet. 38 (9) (2001) 611–614.
[46] R. Mukherjee, et al., A detailed phenotypic description of autosomal dominant cone dystrophy due to a de novo mutation in the GUCY2D gene, Eye (Lond.) 28 (4) (2014) 481–487.
[47] C.L. Tucker, et al., Biochemical analysis of a dimerization domain mutation in RetGC-1 associated with dominant cone-rod dystrophy, Proc. Natl Acad. Sci. U. S.A. 96 (16) (1999) 9039–9044.
[48] S.E. Wilkie, et al., Functional characterization of missense mutations at codon 838 in retinal guanylate cyclase correlates with disease severity in patients with autosomal dominant cone-rod dystrophy, Hum. Mol. Genet. 9 (20) (2000) 3065–3073.
[49] A.M. Dizhoor, E.V. Olshevskaya, I.V. Peshenko, The R838S mutation in Retinal Guanylyl Cyclase 1 (RetGC1) alters calcium sensitivity of cGMP synthesis in the retina and causes blindness in transgenic mice, J. Biol. Chem. 291 (47) (2016) 24504–24516.
[50] S. Sato, et al., GUCY2D Cone-rod dystrophy-6 Is a "phototransduction disease" triggered by abnormal calcium feedback on retinal membrane guanylyl cyclase 1, J. Neurosci. 38 (12) (2018) 2990–3000.
[51] W. Yu, Z. Wu, In vivo applications of CRISPR-based genome editing in the retina, Front. Cell Dev. Biol. 6 (2018) 53.
[52] T.P. Dryja, et al., Mutations within the rhodopsin gene in patients with autosomal dominant retinitis pigmentosa, N. Engl. J. Med. 323 (19) (1990) 1302–1307.
[53] A. Iannaccone, et al., Retinitis pigmentosa associated with rhodopsin mutations: correlation between phenotypic variability and molecular effects, Vis. Res. 46 (27) (2006) 4556–4567.
[54] T.J. McGill, et al., Optomotor and immunohistochemical changes in the juvenile S334ter rat, Exp. Eye Res. 104 (2012) 65–73.
[55] B. Rossmiller, H. Mao, A.S. Lewin, Gene therapy in animal models of autosomal dominant retinitis pigmentosa, Mol. Vis. 18 (2012) 2479–2496.
[56] J. Nishiyama, T. Mikuni, R. Yasuda, Virus-mediated genome editing via homology-directed repair in mitotic and postmitotic cells in mammalian brain, Neuron 96 (4) (2017) 755–768. e5.
[57] N.E. Bengtsson, et al., Muscle-specific CRISPR/Cas9 dystrophin gene editing ameliorates pathophysiology in a mouse model for Duchenne muscular dystrophy, Nat. Commun. 8 (2017) 14454.
[58] W.H. Wu, et al., CRISPR repair reveals causative mutation in a preclinical model of retinitis pigmentosa, Mol. Ther. 24 (8) (2016) 1388–1394.
[59] D.H. Jo, et al., CRISPR-Cas9-mediated therapeutic editing of *Rpe65* ameliorates the disease phenotypes in a mouse model of Leber congenital amaurosis., Sci. Adv. 5 (10) (2019) eaax1210.

CHAPTER

9

The ethics of gene editing in human stem cells

Alangoya Tezel

University of Michigan Medical School, Ann Arbor, MI, United States

Humanity has struggled with the same, unanswered ethical questions since Greek antiquity. Now, in an era of unprecedented technological advancement, we are faced with new ethical conundrums. At the forefront of science and medicine are questions of gene editing. With the potential to forever eradicate diseases and save the lives of millions, the practice of gene editing in stem cells is increasingly under the ethical spotlight. Navigating ethical concerns is a necessary step in unlocking the future of precision medicine. This chapter will review the broader ethical debate surrounding stem cell research and then discuss the specific issues raised by both supporters and opponents of gene editing in stem cells.

Overview of legal and political regulation

In 1998, after nearly 20 years of working with mouse embryonic stem cells, a team of American scientists led by James Thomson successfully isolated and cultured human embryonic stem cells (hESCs) from donated preimplantation embryos [1]. The public reaction was both excitement and fury [2]. On one hand, the clinical implications were tremendous. Within these pluripotent cells lay potential cures for diabetes, AIDS, Parkinson's disease, Alzheimer's disease, and spinal cord injuries. Yet, however, noble the aims of the research, many found the subsequent destruction of preimplantation human embryos to be unjustifiable.

Amid national anxieties, President George W. Bush signed an executive order in 2001 banning the use of federal funds on newly created hESC lines [3]. This decree was an attempt at compromise, curbing the destruction of future embryos while still offering a path, albeit limited, for researchers to continue the development of life-saving therapies. Unfortunately, many of the preexisting lines still approved for federal funding lacked genetic diversity and had already begun accruing genetic mutations. Many researchers opted instead to compete for private funding sources and state-funded initiatives in the effort to create new hESC lines [4].

In 2005, the President's Council on Bioethics presented a new third option: the development of, and federal funding for, "alternative

CRISPR Genome Surgery in Stem Cells and Disease Tissues
DOI: https://doi.org/10.1016/B978-0-12-817876-8.00005-X

sources" of pluripotent stem cells that did not involve the destruction of embryos [5]. Some suggestions included (1) extracting hESCs from cadaveric embryos; (2) nonharmful biopsy of living embryos; (3) creating artificial embryo-like cellular systems; or (4) reprogramming adult somatic cells back to pluripotency [6]. While President Obama revoked most federal funding restrictions for stem cell research in 2009, the Bush administration's hard stance on stem cell research promoted the development of innovative alternatives that otherwise may not have flourished [3]. In 2019, the Trump administration issued a ban on federal funding for research using fetal tissue, citing ethical concerns about the collection of tissue from elective abortion procedures [7]. This ban was later reversed in 2021 under the Biden administration [8]. Many scientists worry that new advancements could place stem cell science on the national radar and usher in another era of federal restrictions [9].

Today, with the advent of gene editing, stem cell science remains entangled in new legal, political, and ethical scrutiny. Understanding the controversy is necessary in order to mitigate concerns, promote advancements, and unleash the full potential of stem cell science.

Human embryonic stem cells

The original controversy surrounding stem cell research was motivated by the same "pro-life" ideology that drives the national debate around abortion. Generally speaking, this view holds that an embryo is a living human being at the moment of fertilization. As such, its destruction is deemed the moral equivalent to murder. The central questions here are whether an embryo possesses human life and whether it is endowed the right to have its life protected.

Some hypothesize that demarcating when life begins relies on understanding when life ends. For instance, it is widely accepted in the medical community that a human being is dead once they no longer exhibit functional brain activity [10]. This event marks the loss of sentience, a complicated notion that can be roughly defined as the capacity to have conscious experiences and react appropriately to external stimuli [11]. At that point, one may be removed from life support and, barring any complications, be eligible for organ donation. By that same logic, the minimum qualification for sentience, and thus life, for an embryo is met 18–25 weeks after conception once brain activity begins [12]. The collection of hESCs for research usually occurs 2 weeks before sentience is thought to arise; therefore harvesting stem cells appears morally permissible because the preimplantation embryo is not yet a "living" being.

Opponents of stem cell research are quick to identify a key difference. While the destruction of a preimplantation embryo does not end actual life, the practice prevents *potential* life. In other words, because embryos have a natural proclivity to develop into a living human being, and so possess the potential for life, they are endowed the human right to have their life protected. As a fertilized egg, embryo, and fetus all have the potential to develop into a human, there is no point in development in which the collection of hESCs is morally permissible.

There are two main criticisms of this potentiality argument. First, it is not apparent that potential rights are the same as actual rights [13]. As a popular analogy, children are potential adults, but they lack many of the same rights as adults, such as the right to vote. Only once a child becomes an adult (legally at age 18 in the United States) will their potential right to vote become an actual right to vote. Along the same line, an embryo is not yet a living human being. Only once it becomes a living human being will it be endowed the right to life. Endowing actual rights to a potential being could be considered a logical leap [11].

Second, supporters of stem cell research have cited empirical data to argue that the existence of potential life does not guarantee actual life. According to conservative estimates, approximately 10–40% of embryos created through intercourse naturally perish before implantation [14], largely as a result of genetic abnormalities. For embryos created using in vitro fertilization (IVF), 15% are lost during thawing and only 14% of those transferred successfully implant [15]. Even if possible, human embryos created through somatic cell nuclear transfer are unlikely to be viable [16]. Some researchers have even created synthetic human embryos with specific alterations—such as the creation of tripronuclear zygotes (egg cells fertilized by two sperm cells)—to prevent any chance for viability [17]. In other words, regardless of the method, the potential for human life is not a guaranteed property of any embryo. At best, embryos possess an abstract "potential to have potential." It is not obvious that this holds the same moral ground.

In response, supporters of the potentiality argument may agree that potential life does not equate to actual life. After all, not every caterpillar will turn into a butterfly, and not every embryo will develop into a human. Yet it is still within reason to assert that stem cell science is an unnecessary obstacle in the development of life. There may be natural limitations for the actualization of life, but that does not justify the introduction of new, avoidable, man-made limitations. As an example, arriving to work on time tomorrow is a possibility that may or may not be actualized. There are various unpredictable factors at play that could cause delay, such as sudden heavy traffic, poor weather conditions, and automobile accidents. Yet, this lack of a guarantee is not an invitation to add even more obstacles, such as stealing one's alarm clock or slashing one's tires.

While a valid concern, there are two logical discrepancies in practice. First, we as a society already artificially block the actualization of potential life. Contraceptive methods, like intrauterine devices and some hormonal contraceptives, prevent pregnancy by blocking fertilization and, if and when that fails, blocking the implantation of the embryo in the uterus by thinning the growth of uterine tissue [18]. This mechanism of action ultimately results in the loss of the 5-day-old preimplantation embryo, an entity that some view as possessing human rights at the moment of fertilization. Surprisingly, these contraceptive methods are not viewed as controversial and are even covered by most health insurance plans. Second, if an embryo were to possess the same moral standing as a human being, it would be common practice to name, mourn, or pray for its loss like one does for the loss of a person, or even perhaps the loss of a fetus (that has, by virtue of being a fetus, established brain activity and can be considered "living"). In fact, it is telling that even colloquially in English, one might refer to an embryo as "it," which is used to signify nonhuman objects, rather than "they," which is used when gender is ambiguous. It is true that the loss of a preimplantation embryo may go unnoticed inside one's body, but many also consent for fertility clinics to actively, and legally, not implant lower viability embryos, effectively blocking their potential for life [19]. These discrepancies do not necessarily discredit the potentiality argument but highlight the inconsistencies in its application.

In reality, the personhood of embryos is often treated in contradictory ways. In the United States, low-income pregnant individuals, regardless of the developmental stage of their child, are treated as one person and allotted only one food stamp [20]. On the other hand, at least 29 states have fetal homicide laws that account for any unborn child in utero [21]. In these states, a homicide involving a pregnant person can be tried as two separate homicides, regardless of the stage of development or viability of the child. The Unborn Victims of Violence Act of 2004 also captures this notion, arguing that a child in utero is "a member of the species *Homo sapiens*, at any stage of development" [22].

Perhaps both sides of the debate can agree that the human embryo has some special status, that natural proclivity for human life, that differentiates it from other excised tissues. As an exercise, consider a fertilized egg cell and a skin cell [23]. Many of us would agree that there is some intangible difference between these two cells. This distinction can create moral and immoral reasons for using embryos. For instance, creating embryos through IVF with the sole intent of monetary gain or cosmetic-toxicology testing would be, in the words of one bioethicist, "disrespectful of the symbolic meaning that many people attach to embryos" [24]. Yet, culturing skin cells for the same cosmetic-toxicology testing seems reasonable. Perhaps collecting excess embryos from successful IVF sessions with the aim of furthering disease-curing research could reaffirm the respect for human life by using the embryo to help preserve human life.

Currently, studies show that almost half of embryos used to help individuals conceive through IVF are discarded during or after the process. While a small proportion is donated to researchers as a result of Obama-era regulations, many states prohibit these embryo donations due to ethical objections. There are arguments for freezing unused embryos, like maintaining their potential for potential life, but disposing of excess embryos does not have a clear moral reason. As President Bush questioned in his 2001 speech—"If embryos are going to be destroyed anyway why shouldn't they be used for the greater good?"

Induced pluripotent stem cells

In the last decade, innovation has shifted the debate. In 2006, a team of Japanese scientists led by Shinya Yamanaka successfully reprogrammed adult mouse fibroblasts into pluripotent stem cells [25]. Using only four transcription factors, researchers could now feasibly harvest their own functional stem cell lines—and better yet, create personalized stem cell lines from an individual's own somatic cells.

The development of these induced pluripotent stem cells (iPSCs) has pushed stem cell science into a new era. Ethical criticisms of the past do not present the same traction because stem cells can now be created without the use of a human embryo. More concretely, the potentiality argument is bypassed by endowing every somatic cell in the body with the potential to generate human life. Consider the fertilized egg cell and the skin cell once again. With the potential to reprogram the skin cell into an egg for subsequent fertilization, the distinction between the two blurs. Moreover, the average human being sheds approximately 40,000 skin cells a minute [26]; to attach a moral standing to every somatic cell in the body would be incoherent.

While iPSCs seem to be a fruitful avenue for future stem cell research, there are still significant hurdles before effective therapies can be enacted. For example, researchers originally believed that creating stem cell lines using a patient's own somatic cells would decrease the risk of immune-system rejection. Yet, recent studies show that iPSCs exhibit an increased risk for tumor formation in vivo [27] and, in some cases [28], increased immunogenicity. Researchers also still struggle with low reprogramming efficiencies, often as a result of using retroviral vectors [29]. Past methods successfully reprogrammed less than 1% of the starting somatic cell population into iPSCs [30,31]. Newer methods report an increased yield of 5–10% of cells—still much lower than what is clinically sufficient [32]. Despite reprogramming, these iPSCs also carry residual DNA methylation signals that are characteristic of their original cell type and bias the iPSCs toward differentiation along their past lineage. These so-called epigenetic memories complicate the generation of alternative cell fates, such as in the effort to create neurons from skin cells to treat neurodegenerative diseases [33]. As the ethical controversy surrounding stem cell science has largely absorbed the entire field, it is necessary to raise public awareness of iPSCs so that researchers can be provided with more funds and resources to solve these technical problems. Otherwise, hESCs, which do not experience any of these reprogramming difficulties, will continue to be the stem cells of choice for researchers, spurring decades of continued ethical debate.

Germline gene editing

Employed only in mice for the past decade [34], gene editing has forever changed the future of disease management and prevention. Gene editing in humans—only attempted in the past few years—remains a contentious issue, especially when conducted on stem cells.

It is first important to pinpoint the exact ethical dilemma raised by gene editing. One important observation is that we as a society already perform gene editing in plants and animals [35,36]. Research suggests that gene editing in crops can raise yields [37], improve quality [38], and accelerate the process of domestication [39,40], offering a possible solution to chronic malnourishment in certain parts of the world and the impending global effects of climate change [41]. As one example, certain genome edits in tomatoes increase the branch number of plants to yield more tomatoes per harvest. Other genome edits can create larger tomatoes with higher levels of important nutrients. While the potential health risks posed by genetically modified crops continues to be debated [42], research involving gene editing in plants is plentiful and growing.

Gene editing in animals is also widely practiced. First and foremost, the creation of mouse models using clustered regularly interspaced short palindromic repeat (CRISPR)/Cas9 nucleases has been an incredibly important tool to study disease [43,44]. Many also believe that gene editing in animals can provide more ethical alternatives to their treatment. For example, 80% of dairy cattle in the United States are physically dehorned every year to protect their handlers [45]. This is a labor-intensive process that is painful for cattle [46]. By identifying naturally existing genetic variants that prevent horn development, researchers are able to successfully edit the genome to produce hornless dairy cattle and minimize the suffering of future offspring.

These examples of gene editing in plants and animals do not mean that there are no ethical issues in their practice; as raised by some ethicists, gene editing has the potential to alter species and ecosystems to cause permanent damage [47]. That being said, the important distinction is that these ethical issues are not at the forefront of the practice. Consider how the Bush-era ethical debate over hESCs spurred such impassioned public discourse that stem cell research was restricted. By contrast, there is no question as to whether gene editing in plants and animals can be used for the greater good and, regulations permitting, the practice is publicly accepted and conducted.

Surprisingly, there have also been instances where gene editing has been deemed ethically acceptable, if not commendable, in children. In 2015, after all conventional treatments failed, Layla, a 1-year-old girl with acute lymphoblastic leukemia, underwent gene therapy [48]. T-cells were extracted from a healthy donor and genetically edited to protect her cells from anticancer medications. After treatment to destroy her original immune cells, the donor T-cells were able to sustain Layla until she received a bone marrow transplant. Eighteen months later, Layla remained healthy, and the researchers in the project were lauded as heroes.

Today, medical research largely focuses on the gene editing of somatic cells, such as blood cells [49] and lung cells [50], as these edits can offer therapeutic benefit to patients like Layla without any effect on future offspring. In contrast, edits made to the germline (gametes and embryo) are passed down from parent to child, such as the edits that led to hornless cattle. For these reasons, the crux of the issue is not the practice of gene editing but rather a specific target of gene editing—most controversially, the human germline.

There are both practical and moral implications to germline gene editing. First, gene editing is still an imprecise science. With a significant risk for off-target effects [51,52], unintended genetic changes as a result of gene editing could be passed down for generations. There have been efforts to mitigate these effects in recent years—such as the delivery of CRISPR through novel methods like circulating red blood cells [53], modified adeno-associated viruses [54,55], or designing variants of the Cas9 RNA-guided DNA endonuclease to reduce nonspecific DNA binding [56]—but many believe further experimentation is necessary before clinical application. There is also the issue of mosaicism, wherein desirable gene edits are only found in a subset of cells [57]. This issue raises questions as to the efficacy of gene editing in curing or preventing diseases.

Second, when making any medical decision that has some associated risk involved, it is necessary to receive consent from those receiving the care. This holds true for both medical treatments and preventative measures, even those with very low risk-to-benefit ratios, such as vaccinations. In the case of germline gene editing, there are two affected parties: the parent and future offspring. The parent may consent, but receiving consent from an embryo to modify its genetic material is impossible. Is this lack of consent, especially in the presence of significant risk, a strong reason to argue that germline editing is immoral?

The answer ties back to the central question of the abortion debate—is an embryo a living being such that it is endowed certain human rights? Yet, one major difference between abortion and germline gene editing is the right, potential or otherwise, that is at stake. In the former, the right to life is at risk because the embryo's potential life is terminated; in the latter, the right to make decisions concerning one's own care is affected because the embryo, which will likely develop into a human being, cannot consent to the gene therapy that could potentially affect its and its potential descendants' qualities of life.

On one hand, the medical community already limits the extent to which pediatric patients can make decisions about their own care. In the United States, patients under 18 years of age cannot make healthcare decisions without parental consent (with the exception of decisions relating to sexual health, mental health, or substance abuse, or patients that are emancipated minors) [58,59]. The laws of other countries may deviate on the so-called age of maturity, but limiting the agency of underage patients is a generally agreed upon practice. Going further than healthcare decisions, couples who pursue preimplantation genetic testing (IVF-PGT) also can make decisions about the genetic code of their offspring by genetically testing zygotes to screen out eggs with a specific sex or risk for certain diseases. Both of these examples suggest that it is not the parental decision-making that is the central moral issue.

The practice of vaccinations in the United States is a helpful analogy. Currently, there are no federal laws that require vaccinations. There are state laws that require certain vaccinations for children attending public schools, but all 50 states offer vaccination exemptions for medical reasons, 48 states for religious reasons, and 20 states for philosophical reasons [60]. Aside from stigma, there is no legal action taken against parents who do not vaccinate their children. As most vaccinations are recommended in the first 15 months of life [61] when the offspring cannot consent, the decision whether to vaccinate relies solely on the parent. In this case, a parent deciding to pursue vaccination for their child, a treatment that will affect the expression of their child's genome by upregulating the production of certain immune cells, is unproblematic. This is in part because research shows that vaccinations are incredibly successful in disease prevention [62,63] and are safe for worldwide use [64,65]. If gene editing could one day properly address issues of specificity and reach the same safety levels, then the parental decision to administer gene editing therapy to prevent disease in a child could be no ethically different than the parental decision to pursue vaccination.

Perhaps one key difference between existing medical treatments and gene editing is that the former only modifies the expression of the genome while the latter permanently alters the DNA sequence. Epigenetics research has shown that various environmental factors—such as diet, exercise, stress, and so on—impact the expression of different parts of the genome [66]. Many drugs also work by upregulating or downregulating certain genes. With ever-changing gene expression, is there any special reason to be attached to the individual sequence of our individual DNA—a sequence that naturally accumulates mutations with every cell replication cycle? A subset of scholars might assert that the genetic code is written by God and its manipulation is strictly prohibited. Yet, an argument that allows for drugs that induce changes in gene expression but condemns the practice of gene editing remains unconvincing. It seems the ethical debate of germline gene editing can be traced back to the target once more—just what edits are we making?

To date, there has only ever been one reported instance of germline editing in living humans. On November 2018, despite the various international bans on germline editing, the world's first genetically edited babies were born. Despite being healthy, the twin girls—Lulu and Nana—incited a global uproar [67]. The project, led by Chinese scientist Jiankui He, used CRISPR/Cas9 to create a delta 32 mutation in the CCR5 gene, a known genetic variant that confers resistance to the human immunodeficiency virus (HIV) [68]. To better test resistance, embryos were selected from HIV-positive fathers and HIV-negative mothers. From He's perspective, Lulu and Nana paved the way for a world free from HIV infection, and given enough momentum, free from other diseases potentially eradicated through germline gene editing. Yet for many others, He's research, conducted largely in secret through misleading consent forms and against the rules of many regulatory bodies, set back the field of gene editing by negatively impacting the public perspective. Moreover, certain aspects of He's methodology were problematic, such as his decision to only edit one copy of the CCR5 gene in one of the twins, an edit that would not confer HIV resistance even if successful [69]. This, combined with the fact that both twins exhibited mosaicism with only 85% of their cells containing the edits [69], add to concerns that the experiment was prematurely and irresponsibly conducted. Lulu and Nana were not in any immediate danger before; was germline gene editing the only way to prevent HIV infection in the embryos or were there other, safer means that did not require forever altering the lives of Lulu, Nana, and their future offspring?

Many researchers would agree that there are safer and more effective alternatives to prevent HIV infection. For this reason, germline gene editing experiments today largely focus on diseases with unmet medical needs, patients like Layla who have already exhausted all other options. As one example, a team led by Dieter Egli, a developmental biologist at Columbia University, uses CRISPR to edit embryonic DNA and prevent mutations that cause retinitis pigmentosa, a group of rare genetic disorders that lead to irreversible blindness [70]. While Egli plans to continue working strictly with human cells, many in the scientific community are anxious about whether the research could, through Egli's lab or another's, be conducted in humans later on.

Another challenge raised by germline editing is the difference between treatment and enhancement. Today, there is no cure for HIV infection [71]. As such, using gene editing to eradicate HIV infection seems to be a universally reasonable target (though more reasonable in those whom have already acquired HIV). Along that same line, using gene editing to prevent incurable diseases such as Huntington's disease, amyotrophic lateral sclerosis, or retinitis pigmentosa like Egli's lab, also seems to be a noble aim. By contrast, using gene editing to prevent baldness or to make one's eyelashes longer, changes that would enhance oneself without treating any clear disease (and are possible to "treat" through alternative methods such as hairpieces or eyelash extensions), seem less fundamental. The distinction between these two ends of the spectrum is clear, but how does one draw the line in gray areas? For example, many individuals with autism can lead happy lives; would gene editing targeted at preventing autism be a treatment or an enhancement? The way in which we define health, disability, and disease will shape the ways gene editing is used in the future. In a world already stigmatizing of those with disabilities, how could gene editing in these gray areas affect our concept of health?

Looking ahead: the slippery slope

Many of the anxieties surrounding stem cell research today focus more on its future implications than present applications. In other words, we may agree that protecting others from HIV infection is a good thing to do in theory, yet we may disagree on whether it is ethical given its future implications—that germline editing become more commonplace and extend past just preventing disease and move toward human enhancement. This phenomenon is succinctly called the slippery-slope argument.

Emboldened perhaps by dystopic science fiction, slippery-slope theorists assert that allowing stem cell research and germline gene editing today (which we can call A) will inevitably lead to neo-eugenics, designer children, human–animal hybridization, and reproductive cloning in the future (B). Logically, this takes a simple structure: if A, then B. The central question here is not whether B will come from A, as there is no empirical way to predict the future. Rather, the focus should be—is fear of B a good enough reason not to do A?

On one hand, there are many regulations already in place to prevent B. In the United States, there are federal bans in place to restrict human reproductive cloning, such as the bipartisan Human Cloning Ban and Stem Cell Protection Act of 2003 introduced by Senators Orrin Hatch (R–UT) and Dianne Feinstein (D–CA) [72]. Moreover, the National Institute of Health restricts funding for research involving germline gene editing, and clinics nationwide are banned from editing embryonic genes. Given the differences in cultural viewpoints, different countries have their own restrictions, such as Brazil's requirement that excess IVF embryos have been frozen for 3 years before donation to research or Singapore's ban on experimenting with embryos older than 2 weeks [73]. In other words, ethical discussion has the power to shape regulations, which in turn can limit certain possibilities. Under this perspective, the fear of B is how we work to make regulations that will prevent B. Thus the fear of B allows A to safely progress.

Another argument is whether fear of B is legitimate. In other words, even if science could one day advance enough to achieve B, it is questionable whether B would be beneficial. Reproductive cloning in humans, if developed, could lead to mass overpopulation and an increasing environmental burden on the planet. Creating designer children risks eliminating the genetic assortment necessary to survive under new environmental pressures. Barring the regulations already preventing these developments, it is unclear whether the scientific community would want to pursue them in the first place. In this case, A could lead to the capability for B but not necessarily the actualization of B.

Yet, as we have seen in the past, sometimes individuals disagree with regulations and pursue research that is not desirable. The most recent example is the controversy surrounding the germline gene editing of Lulu and Nana, in which national regulations were disregarded and clinical trials run largely in secret. As scientific development affects all of humankind, the international community must work together to determine the limits and scope of research to prevent rogue experimentation [74].

If anything, the ethical debates of the past two millennia have taught us that scientific development is inevitable. When faced with restrictions concerning hESCs, researchers developed iPSCs. After failing to treat Layla's leukemia, researchers implemented experimental gene therapy. Scientific research is constantly rising to the occasion, addressing new challenges and raising new questions. When moving forward, caution—not prevention—is the answer.

References

[1] J.A. Thomson, J. Itskovitz-Eldor, S.S. Shapiro, M.A. Waknitz, J.J. Swiergiel, V.S. Marshall, et al., Embryonic stem cell lines derived from human blastocysts, Science 282 (5391) (1998) 1145–1147.

[2] M.C. Nisbet, Public opinion about stem cell research and human cloning, Public Opin. Q. 68 (1) (2004) 131–154.

[3] V. Murugan, Stem cell issue: embryonic stem cell research: a decade of debate from Bush to Obama, Yale J. Biol. Med. 82 (3) (2009) 101.

[4] I. Hyun, The bioethics of stem cell research and therapy, J. Clin. Invest. 120 (1) (2010) 71–75.

[5] A. Schulman, The search for alternative sources of human pluripotent stem cells, Stem Cell Rev. Rep. 1 (4) (2005) 291–292.

[6] The President's Council on Bioethics, White Paper: Alternative Sources of Human Pluripotent Stem Cells, National Academic Press, Washington, DC, 2005.

[7] Sara Reardon, Trump administration halts fetal-tissue research by government scientists, Nature 570 (7760) (2019) 148–149.

[8] Nidhi Subbaraman, NIH reverses Trump-era restrictions on fetal-tissue research, Nature (2021).

[9] E. Mullin, Under Trump, biologists fear political risks of controversial research. <https://www.technologyreview.com/s/609323/under-trump-biologists-seek-a-low-profile-for-controversial-research/>, 2017 (accessed 10.11.17).

[10] E.F. Wijdicks, The diagnosis of brain death, N. Engl. J. Med. 344 (16) (2001) 1215–1221.

[11] L. Bortolotti, J. Harris, Stem cell research, personhood and sentience, Reprod. Biomed. Online 10 (2005) 68–75.

[12] S. Tawia, When is the capacity for sentience acquired during human fetal development? J. Maternal-Fetal Med. 1 (3) (1992) 153–165.

[13] P. Singer, Practical Ethics, Cambridge University Press, 1993, p. 153.

[14] G.E. Jarvis, Early embryo mortality in natural human reproduction: What the data say, F1000Research 5 (2016) 2765.

[15] M.E. Pavone, J. Innes, J. Hirshfeld-Cytron, R. Kazer, J. Zhang, Comparing thaw survival, implantation and live birth rates from cryopreserved zygotes, embryos and blastocysts, J. Hum. Reprod. Sci. 4 (1) (2011) 23.

[16] S.M. Rhind, J.E. Taylor, P.A. De Sousa, T.J. King, M. McGarry, I. Wilmut, Human cloning: can it be made safe? Nat. Rev. Genet. 4 (11) (2003) 855.

[17] P. Liang, Y. Xu, X. Zhang, C. Ding, R. Huang, Z. Zhang, et al., CRISPR/Cas9-mediated gene editing in human tripronuclear zygotes, Protein Cell 6 (5) (2015) 363–372.

[18] J.A. Spinnato II, Mechanism of action of intrauterine contraceptive devices and its relation to informed consent, Am. J. Obstet. Gynecol. 176 (3) (1997) 503–506.

[19] G. Yao, J. Xu, Z. Xin, W. Niu, S. Shi, H. Jin, et al., Developmental potential of clinically discarded human embryos and associated chromosomal analysis, Sci. Rep. 6 (2016) 23995.

[20] T. Pavlenko, Many low-income pregnant women don't have enough to eat, despite food programs. <https://www.centerforhealthjournalism.org/2018/05/31/many-low-income-pregnant-women-dont-get-enough-eat-despite-federal-food-programs>, 2018.

[21] A.S. Murphy, A survey of state fetal homicide laws and their potential applicability to pregnant women who harm their own fetuses, Ind. LJ 89 (2014) 847.

[22] M. Fitzpatrick, Fetal personhood after the Unborn Victims of Violence Act, Rutgers L. Rev. 58 (2) (2005) 553.

[23] B.A. Manninen, Revisiting the argument from fetal potential, Philos. Ethics Humanit. Med. 2 (1) (2007) 7.

[24] J.A. Robertson, Human embryonic stem cell research: ethical and legal issues, Nat. Rev. Genet. 2 (1) (2001) 74.
[25] K. Takahashi, S. Yamanaka, Induction of pluripotent stem cells from mouse embryonic and adult fibroblast cultures by defined factors, Cell 126 (4) (2006) 663–676.
[26] J.A. Peterson, Ten things people ought to know about skin, ACSM's Health & Fit. J. 22 (2) (2018) 50.
[27] M. Reisman, K.T. Adams, Stem cell therapy: a look at current research, regulations, and remaining hurdles, Pharm. Therapeut. 39 (12) (2014) 846.
[28] T. Zhao, Z.N. Zhang, Z. Rong, Y. Xu, Immunogenicity of induced pluripotent stem cells, Nature 474 (7350) (2011) 212.
[29] K. Okita, T. Ichisaka, S. Yamanaka, Generation of germline-competent induced pluripotent stem cells, Nature 448 (7151) (2007) 313.
[30] T. Brambrink, R. Foreman, G.G. Welstead, C.J. Lengner, M. Wernig, H. Suh, et al., Sequential expression of pluripotency markers during direct reprogramming of mouse somatic cells, Cell Stem Cell 2 (2) (2008) 151–159.
[31] I.H. Park, R. Zhao, J.A. West, A. Yabuuchi, H. Huo, T. A. Ince, et al., Reprogramming of human somatic cells to pluripotency with defined factors, Nature 451 (7175) (2008) 141.
[32] J.M. Polo, E. Anderssen, R.M. Walsh, B.A. Schwarz, C. M. Nefzger, S.M. Lim, et al., A molecular roadmap of reprogramming somatic cells into iPS cells, Cell 151 (7) (2012) 1617–1632.
[33] K. Kim, A. Doi, B. Wen, K. Ng, R. Zhao, P. Cahan, et al., Epigenetic memory in induced pluripotent stem cells, Nature 467 (7313) (2010) 285.
[34] E.E. Perez, J. Wang, J.C. Miller, Y. Jouvenot, K.A. Kim, O. Liu, et al., Establishment of HIV-1 resistance in CD4+ T cells by genome editing using zinc-finger nucleases, Nat. Biotechnol. 26 (7) (2008) 808.
[35] D. Carroll, R.A. Charo, The societal opportunities and challenges of genome editing, Genome Biol. 16 (1) (2015) 242.
[36] D. Carroll, A.L. Van Eenennaam, J.F. Taylor, J. Seger, D.F. Voytas, Regulate genome-edited products, not genome editing itself, Nat. Biotechnol. 34 (5) (2016) 477.
[37] J. Poland, J. Endelman, J. Dawson, J. Rutkoski, S. Wu, Y. Manes, et al., Genomic selection in wheat breeding using genotyping-by-sequencing, Plant Genome 5 (3) (2012) 103–113.
[38] A. Zsögön, T. Čermák, E.R. Naves, M.M. Notini, K.H. Edel, S. Weinl, et al., De novo domestication of wild tomato using genome editing, Nat. Biotechnol. 36 (2018) 1211–1216.
[39] T. Li, X. Yang, Y. Yu, X. Si, X. Zhai, H. Zhang, et al., Domestication of wild tomato is accelerated by genome editing, Nat. Biotechnol. 36 (2018) 1160–1163.
[40] Z.H. Lemmon, N.T. Reem, J. Dalrymple, S. Soyk, K.E. Swartwood, D. Rodriguez-Leal, et al., Rapid improvement of domestication traits in an orphan crop by genome editing, Nat. Plants 4 (10) (2018) 766.
[41] Y. Zhang, K. Massel, I.D. Godwin, C. Gao, Applications and potential of genome editing in crop improvement, Genome Biol. 19 (1) (2018) 210.
[42] A. Hilbeck, R. Binimelis, N. Defarge, R. Steinbrecher, A. Székács, F. Wickson, et al., No scientific consensus on GMO safety, Environ. Sci. Europe 27 (1) (2015) 4.
[43] D. Heckl, M.S. Kowalczyk, D. Yudovich, R. Belizaire, R.V. Puram, M.E. McConkey, et al., Generation of mouse models of myeloid malignancy with combinatorial genetic lesions using CRISPR-Cas9 genome editing, Nat. Biotechnol. 32 (9) (2014) 941.
[44] H. Mou, Z. Kennedy, D.G. Anderson, H. Yin, W. Xue, Precision cancer mouse models through genome editing with CRISPR-Cas9, Genome Med. 7 (1) (2015) 53.
[45] D.F. Carlson, C.A. Lancto, B. Zang, E.S. Kim, M. Walton, D. Oldeschulte, et al., Production of hornless dairy cattle from genome-edited cell lines, Nat. Biotechnol. 34 (5) (2016) 479.
[46] P.M. Faulkner, D.M. Weary, Reducing pain after dehorning in dairy calves, J. Dairy Sci. 83 (9) (2000) 2037–2041.
[47] J. Lunshof, Regulate gene editing in wild animals, Nature 521 (7551) (2015).
[48] S. Reardon, Leukaemia success heralds wave of gene-editing therapies, Nature 527 (7577) (2015) 146.
[49] P. Tebas, D. Stein, W.W. Tang, I. Frank, S.Q. Wang, G. Lee, et al., Gene editing of CCR5 in autologous CD4 T cells of persons infected with HIV, N. Engl. J. Med. 370 (10) (2014) 901–910.
[50] S.G. Swisher, J.A. Roth, J. Nemunaitis, D.D. Lawrence, B.L. Kemp, C.H. Carrasco, et al., Adenovirus-mediated p53 gene transfer in advanced non-small-cell lung cancer, J. Natl Cancer Inst. 91 (9) (1999) 763–771.
[51] Y. Fu, J.A. Foden, C. Khayter, M.L. Maeder, D. Reyon, J.K. Joung, et al., High-frequency off-target mutagenesis induced by CRISPR-Cas nucleases in human cells, Nat. Biotechnol. 31 (9) (2013) 822.
[52] T.J. Cradick, E.J. Fine, C.J. Antico, G. Bao, CRISPR/Cas9 systems targeting β-globin and CCR5 genes have substantial off-target activity, Nucleic Acids Res. 41 (20) (2013) 9584–9592.
[53] J. Shi, L. Kundrat, N. Pishesha, A. Bilate, C. Theile, T. Maruyama, et al., Engineered red blood cells as carriers for systemic delivery of a wide array of functional probes, Proc. Natl Acad. Sci. U.S.A. 111 (28) (2014) 10131–10136.

[54] W.L. Chew, M. Tabebordbar, J.K. Cheng, P. Mali, E.Y. Wu, A.H. Ng, et al., A multifunctional AAV–CRISPR–Cas9 and its host response, Nat. Methods 13 (10) (2016) 868.

[55] E. Senís, C. Fatouros, S. Große, E. Wiedtke, D. Niopek, A.K. Mueller, et al., CRISPR/Cas9-mediated genome engineering: an adeno-associated viral (AAV) vector toolbox, Biotechnol. J. 9 (11) (2014) 1402–1412.

[56] B.P. Kleinstiver, V. Pattanayak, M.S. Prew, S.Q. Tsai, N.T. Nguyen, Z. Zheng, et al., High-fidelity CRISPR–Cas9 nucleases with no detectable genome-wide off-target effects, Nature 529 (7587) (2016) 490.

[57] M. Mehravar, A. Shirazi, M. Nazari, M. Banan, Mosaicism in CRISPR/Cas9-mediated genome editing, Dev. Biol. 445 (2) (2018) 156–162.

[58] Committee on Bioethics, Informed consent, parental permission, and assent in pediatric practice, Pediatrics 95 (2) (1995) 314–317.

[59] H. Boonstra, E. Nash, Minors and the right to consent to health care, Guttmacher Rep. Public Policy 3 (4) (2000) 4–8.

[60] D.A. Salmon, S.P. Teret, C.R. MacIntyre, D. Salisbury, M.A. Burgess, N.A. Halsey, Compulsory vaccination and conscientious or philosophical exemptions: past, present, and future, Lancet 367 (9508) (2006) 436–442.

[61] Immunization Schedules. <https://www.cdc.gov/vaccines/schedules/hcp/imz/child-adolescent.html>, 2018.

[62] P.A. Gross, A.W. Hermogenes, H.S. Sacks, J. Lau, R.A. Levandowski, The efficacy of influenza vaccine in elderly persons: a meta-analysis and review of the literature, Ann. Intern. Med. 123 (7) (1995) 518–527.

[63] R. Biellik, S. Madema, A. Taole, A. Kutsulukuta, E. Allies, R. Eggers, et al., First 5 years of measles elimination in southern Africa: 1996–2000, Lancet 359 (9317) (2002) 1564–1568.

[64] P.I. Folb, E. Bernatowska, R. Chen, J. Clemens, A.N. Dodoo, S.S. Ellenberg, et al., A global perspective on vaccine safety and public health: the Global Advisory Committee on Vaccine Safety, Am. J. Public Health 94 (11) (2004) 1926–1931.

[65] W. Zhou, V. Pool, J.K. Iskander, R. English-Bullard, R. Ball, R.P. Wise, et al., Surveillance for safety after immunization: vaccine adverse event reporting system (VAERS)—United States, 1991–2001, MMWR Surveill. Summ. 52 (1) (2003) 1–24.

[66] A. Bird, Perceptions of epigenetics, Nature 447 (7143) (2007) 396.

[67] A. Gouw, The CRISPR advent of Lulu and Nana, Theol. Sci. 17 (1) (2019) 9–12.

[68] G. Hütter, D. Nowak, M. Mossner, S. Ganepola, A. Müßig, K. Allers, et al., Long-term control of HIV by CCR5 Delta32/Delta32 stem-cell transplantation, N. Engl. J. Med. 360 (7) (2009) 692–698.

[69] O. Dyer, Researcher who edited babies' genome retreats from view as criticism mounts, BMJ 363 (2018) k5113.

[70] R. Stein, New United States experiments aim to create gene-edited human embryos. <https://www.npr.org/sections/health-shots/2019/02/01/689623550/new-u-s-experiments-aim-to-create-gene-edited-human-embryos>, 2019.

[71] S.G. Deeks, B. Autran, B. Berkhout, M. Benkirane, S. Cairns, N. Chomont, et al., Towards an HIV cure: a global scientific strategy, Nat. Rev. Immunol. 12 (8) (2012) 607.

[72] G.D. Fischbach, R.L. Fischbach, Stem cells: science, policy, and ethics, J. Clin. Invest. 114 (10) (2004) 1364–1370.

[73] D. Dhar, J.H. Hsi-En, Stem cell research policies around the world, Yale J. Biol. Med. 82 (3) (2009) 113–115.

[74] D. Pei, D.W. Beier, E. Levy-Lahad, G. Marchant, J. Rossant, J.C.I. Belmonte, et al., Human embryo editing: opportunities and importance of transnational cooperation, Cell Stem Cell 21 (4) (2017) 423–426.

Index

Note: Page numbers followed by "*f*" and "*t*" refer to figures and tables, respectively.

Kolibri 360 GmbH
Ferdinand-Jühlke-Straße 7
99095 Erfurt, Deutschland
produktsicherheit@kolibri360.de